Guides to Information Sources

Information Sources in the

Life Sciences

Fourth Edition

Guides to Information Sources

A series under the General Editorship of
Ia C. McIlwaine,
M.W. Hill and
Nancy J. Williamson

This series was known previously as 'Butterworths Guides to Information Sources'.

Other titles available include:

Information Sources in Official Publications
edited by Valerie J. Nurcombe
Information Sources in Law
edited by Jules Winterton and Elizabeth M. Moys
Information Sources in Architecture and Construction (Second edition)
edited by Valerie J. Nurcombe
Information Sources in Chemistry (Fourth edition)
edited by R.T. Bottle and J.F.B. Rowland
Information Sources in Physics (Third edition)
edited by Dennis F. Shaw
Information Sources in Finance and Banking
by Ray Lester
Information Sources in Environmental Protection
edited by Selwyn Eagle and Judith Deschamps
Information Sources in Grey Literature (Third edition)
by C.P. Auger
Information Sources in Music
edited by Lewis Foreman
Information Sources in Sport and Leisure
edited by Michele Shoebridge
Information Sources in Patents
edited by C.P. Auger
Information Sources for the Press and Broadcast Media
edited by Selwyn Eagle
Information Sources in Information Technology
edited by David Haynes
Information Sources in Pharmaceuticals
edited by W.R. Pickering
Information Sources in Metallic Materials
edited by M.N. Patten
Information Sources in the Earth Sciences (Second edition)
edited by David N. Wood, Joan E. Hardy and Anthony P. Harvey
Information Sources in Cartography
edited by C.R. Perkins and R.B. Barry
Information Sources in Polymers and Plastics
edited by R.T. Adkins
Information Sources in Economics (Second edition)
edited by John Fletcher

Guides to Information Sources

Information Sources in the
Life Sciences
Fourth Edition

Editor

H.V. Wyatt

BOWKER
SAUR

London · Melbourne · Munich · New Providence, NJ

British Library Cataloguing in Publication Data
Information sources in life sciences. – 4th ed.
1. Life sciences – Bibliography 2. Life sciences – Information services
I. Wyatt, H. V. (Harold Vivian), 1926–
570.7

ISBN 1857390709

Library of Congress Cataloging-in-Publication Data
Information sources in life sciences/editors, H.V. Wyatt. – 4th ed.
p. cm. – (Guides to information sources)
Rev. ed. of: Information sources in the life sciences. 3rd ed. 1987.
Includes bibliographical references and index.
ISBN 1–85739–070–9 (alk. paper)
1. Life sciences literature. I. Wyatt, H. V. (Harold Vivian)
II. Information sources in the life sciences. III. Series: Guides to information sources (London, England)
QH303.6.I54 1997
016.57–dc21 96–29868
CIP

Published by Bowker-Saur, Maypole House, Maypole Road.
East Grinstead, West Sussex RH19 1HU, UK
Tel: +44(0)1342 330100 Fax: +44(0)1342 330191
E-mail: lis@bowker-saur.com
Internet Website: http://www.bowker-saur.com/service/
Bowker-Saur is part of REED BUSINESS INFORMATION

ISBN 1–85739–070–9

Cover design by Calverts Press
Phototypeset by Intype London Ltd
Printed on acid-free paper
Printed and bound in Great Britain by Antony Rowe Ltd, Chippenham

Series editors' foreword

The second half of the 20th century has been characterized by the recognition that our style of life depends on acquiring and using information effectively. It has always been so, but only in the information society has the extent of the dependence been recognized and the development of technologies for handling information become a priority. These modern technologies enable us to store more information, to select and process parts of the store more skilfully and transmit the product more rapidly than we would have dreamt possible only 40 years ago. Yet the irony still exists that, while we are able to do all this and are assailed from all sides by great masses of information, ensuring that one has what one needs just when one wants it is frequently just as difficult as ever. Knowledge may, as Johnson said in the well known quotation, be of two kinds, but information, in contrast, is of many kinds and most of it is, for each individual, knowable only after much patient searching.

The aim of each Guide in this series is simple. It is to reduce the time which needs to be spent on that patient searching; to recommend the best starting point and sources most likely to yield the desired information. Like all subject guides, the sources discussed have had to be selected, and the criteria for selection will be given by the individual editors and will differ from subject to subject. However, the overall objective is constant; that of providing a way into a subject to those new to the field or to identify major new or possibly unexplored sources to those already familiar with it.

The great increase in new sources of information and the overwhelming input of new information from the media, advertising, meetings and conferences, letters, internal reports, office memoranda, magazines, junk mail, electronic mail, fax, bulletin boards etc. inevitably tend to make one reluctant to add to the load on the mind and memory by consulting books and journals. Yet they, and the other traditional types of printed material, remain for many purposes the most reliable sources of information. Despite all the information that is instantly accessible via the new technologies one still has to look things up in databooks, monographs, journals, patent specifications, standards, reports both official and commercial, and on maps and in atlases. Permanent recording of facts, theories and opinions is still carried out primarily by publishing in printed form. Musicians still work from printed scores even though they are helped by sound recordings. Sailors still use printed charts and tide tables even though they have radar and sonar equipment.

However, thanks to computerized indexes, online and CD-ROM, searching the huge bulk of technical literature to draw up a list of references can be undertaken reasonably quickly. The result, all too often, can still be a formidably long list, of which a knowledge of the nature and structure of information sources in that field can be used to put it in order of likely value.

It is rarely necessary to consult everything that has been published on the topic of a search. When attempting to prove that an invention is genuinely novel, a complete search may seem necessary, but even then it is common to search only obvious sources and leave it to anyone wishing to oppose the grant of a patent to bear the cost of hunting for a prior disclosure in some obscure journal. Usually, much proves to be irrelevant to the particular aspect of our interest and whatever is relevant may be unsound. Some publications are sadly lacking in important detail and present broad generalizations flimsily bridged with arches of waffle. In any academic field there is a 'pecking order' of journals so that articles in one journal may be assumed to be of a higher or lower calibre than those in another. Those experienced in the field know these things. The research scientist soon learns, as it is part of his training, the degree of reliance he can place on information from co-workers elsewhere, on reports of research by new and (to him) unknown researchers on data compilations and on manufacturers of equipment. The information worker, particularly when working in a field other than his own, faces very serious problems as he tries to compile, probably from several sources, a report on which his client may base important actions. Even the librarian, faced only with recommending two or three books or journal articles, meets the same problem though less acutely.

In the Bowker-Saur Guides to Information Sources we aim to bring you the knowledge and experience of specialists in the field. Each author regularly uses the information sources and services described and any tricks of the trade that the author has learnt are passed on.

Nowadays, two major problems face those who are embarking upon research or who are in charge of collections of information of every kind. One is the increasingly specialized knowledge of the user and the concomitant ignorance of other potentially useful disciplines. The second problem is the trend towards cross-disciplinary studies. This has led to a great mixing of academic programmes – and a number of imprecisely defined fields of study. Courses are offered in Environmental Studies, Women's Studies, Communication Studies or Area Studies, and these are the forcing ground for research. The editors are only too aware of the difficulties raised by those requiring information from such hybrid subject fields and this approach, too, is being handled in the series alongside the traditional 'hard disciplines'.

Guides to the literature have a long and honoured history. Marion Spicer of SRIS recently drew to our attention a guide written in 1891 for engineers. No doubt there are even earlier ones. Nowadays, with the information and even the publishing fields changing quite frequently, it is necessary to update guides every few years and this we do in this present Series.

Michael Hill
Ia McIlwaine
Nancy Williamson

About the contributors

M. Burdon, Ph.D., was a post-doctoral fellow at the Syntex Institute for Molecular Biology in California after reading chemistry at Cambridge. He continued research at the Universities of Warwick and Edinburgh until moving to the University of St. Andrews, Scotland where he is Lecturer in Biochemistry. His research interests have included bacteriophage genetics, cystic fibrosis and the molecular biology of mycoplasmas.
mgb2@st-and.ac.uk

Angela Hibbert, B.Sc., graduated in Biological Sciences from Portsmouth Polytechnic. She joined BIOSIS UK, York and is Group Leader in the User Services Department. She maintains the manuals and vocabularies used in the Zoological Record products, prepares and reviews user documentation and represents BIOSIS at exhibitions.
ahibbert@york.biosis.org

P. J. Large, D.Sc., F.I. Biol., is Reader in Microbial Physiology in the Department of Biological Sciences at the University of Hull. He is a graduate of Sheffield, Oxford and Hull Universities and author of two books, chapters in student texts and many research papers in biochemistry, biotechnology and microbiology. He is a member of the editorial board of *Microbiology* and an Associate of the Institute of Linguists. He has interests in beer, computers and the culture and language of The Netherlands and Flanders.
p.j.large@biosci.hull.ac.uk

R. Muller, D.Sc, F.I. Biol. was Lecturer at Ibadan University in Nigeria and then Senior Lecturer at the London School of Hygiene

and Tropical Medicine. He was lately Director of the International Institute of Parasitology. He was President of the European Federation of Parasitologists 1988–1992, edits two journals of parasitology and is Honorary General Secretary of the British Society for Parasitology. He has published three books.
101317.3253@compuserve.com

Helen Peat, D.Phil., a graduate in botany, took the M.Sc. Information Sciences at Sheffield University and then compiled a database on the ecology of the British flora for a D.Phil at York University. She is currently the Data and Resource Centre Manager in the Terrestrial and Freshwater Life Sciences Division of the British Antarctic Survey at Cambridge.
h.peat@bas.ac.uk

N. J. Robinson, B.Sc., M.Inf.Sci., graduated in zoology from Reading University and became an indexer for BIOSIS UK, York, producing the *Zoological Record* database. He is now User Services Manager in charge of the European Help Desk and UK training on all BIOSIS products and produces user documentation. His department maintains, develops and implements the controlled vocabularies for the *Zoological Record*.
nrobinson@york.biosis.org

A.D. Smith, Ph.D., was a Commonwealth Fellow at the Albert Einstein College of Medicine in New York from 1963–1964, returning to become Lecturer and then Reader at Middlesex Hospital Medical School. He retired in 1991 to be Emeritus Reader in Biochemistry and Honorary Research Associate in the Department of Chemical Pathology, University College, London.
a.smith@ucl.ac.uk

H.V. Wyatt, Ph.D., F.I. Biol., was a marine zoologist before changing to microbiology. He has held posts at St. Bartholomew's Hospital, Johns Hopkins Medical Institutions, Leeds and Manchester Universities, National Cancer Institute and in the West Bank and India. He was Reader in Microbiology at Bradford University until closure of the department. He is now Honorary Research Fellow in Public Health Medicine, School of Healthcare Studies, University of Leeds and has over 250 publications.
nurhvw@leeds.ac.uk

T. D. Wyatt, Ph.D., is Research Associate in the Department of Zoology and University Lecturer in Biological Sciences in the Department for Continuing Education, Oxford University where he teaches and organises courses for the general public and professional scientists. He is a Fellow of Kellogg College. Since research at Cambridge on the ecology and behaviour of a subsocial

intertidal beetle, he has held teaching and research posts at Leeds and Cardiff. His research is focused on insect pheromones and behaviour.
wyatt@vax.ox.ac.uk

Contents

INTRODUCTION

H.V. WYATT

In the last decade, biology and information have been transformed and so has this book: it has been completely rewritten. The techniques of molecular biology, immunology, genetics and biochemistry have invaded all subjects, including behaviour (p. 227) so that scientists may have more in common with scientists using the same techniques in different subjects, than with their colleagues. While information scientists deplore the proliferation, specialist journals arise because scientists need them. Library budgets have contracted although there has been an explosion in data generated by the new techniques. University library budgets are being cut and even national resources are under threat: five thousand medical journals are to be cut at SRIS (p. 98).

Nevertheless, new electronic networks are available, although addresses change rapidly and new appear. International, national and specialist societies can maintain and distribute addresses through their journals and newsletters. Research newsletters (p. 27), libraries and information scientists can keep scientists informed. The health satellite, SatelLife, which links resources in Boston with third world hospitals (see p. 97) now has links in twenty two African, five Asian including China and two Latin American countries (Groves, 1996). How long before there is a similar biological satellite link?

The cheap and easy access to journal literature through electronic links (see for example BIDS, p. 69) has meant that older literature not on CD-ROM or in databases is rapidly forgotten. Thus, sexual transmission of Brucellosis to monkeys was shown in 1906, but has been forgotten (Wyatt, 1996). Other literature, from 'grey' sources, books, newspapers, annual reports etc may be necessary for a complex subject (see the example of *Rabies* p. 71).

Modern research is very expensive: governments and business expect returns for the investment. Yet the campaigns to eradicate Chagas' disease and the guinea worm *Dracunculus* (p. 195) are based on the simple application of known techniques. The eradication of poliomyelitis is based on the Sabin vaccine of the 1950s (see p. 194). The most effective means yet to cut malaria infections is the use of impregnated bednets (p. 199). While the most modern techniques are used to study malaria (see for example p. 189), a recent review (Anderson et al, 1996) lists 39 examples of promising research which have not been developed further and the authors comment.

> many areas of basic biology and biochemistry remain unexplored as funding over the past 10–15 years has been directed towards vaccine development.

Applying knowledge can be far more difficult than acquiring new. Exciting new research gains high Impact Factors (see Chapter 10), but application is more difficult to finance, publish and trace.

We hope that guides to information sources, like this book, will be read by library users whenever they have a query. Books on writing and speaking are also intended for all scientists. Such books are a special resource and should be prominently displayed for users to browse: in libraries where they have been displayed, students and staff have used them. More scientists should evaluate the usefulness of sources and tools (see for example p. 55). As CD-ROMs and the net become available, it is interesting that two groups prefer the printed versions (see pp. 20, 29).

References

Anderson J, MacLean M, Davies C. *Malaria research. An audit of international activity.* PRISM Report no. 7. London, The Wellcome Trust 1996.

Groves T. SatelLife: getting relevant information to the developing world. *BMJ* 1996; 313: 1606–1609 (Christmas Issue).

Wyatt H.V. *Brucella melitensis* can be transmitted sexually. *Lancet* 1996; 348: 615.

CHAPTER ONE

Reading for profit: current awareness

H.V. WYATT

Experimental scientists must be aware of what other scientists are doing. Scientists in the more traditional disciplines and ecology are more dependent on past literature and the specimens in museum collections. The life sciences overlap with both medicine and agriculture and their literatures, as well as foods and the environment. The volume of literature and information required is therefore greater than in many other science subjects. This book seeks to guide readers towards sources rather than to be comprehensive.

> *News is either important or unimportant. If it is unimportant, I don't want to know about it. If it is important, someone is sure to tell me about it.* (Arthur Balfour, UK Prime Minister 1902–1905)

Science is a social activity in which we compete with our peers. We have to publish to keep our place, obtain new grants and receive invitations to speak, referee and edit. We need to read what others have published, to avoid repetition and to help us decide on the next experiment. We are members of one or more small invisible colleges, each of which, regardless of the total number of journals, publishes in a few major, high-profile journals and a small group of more specialist journals. Scientists want to publish, but only read reluctantly. Scientists usually want to work at the bench, or supervise and plan experiments. Reading the literature is a necessary distraction. Older scientists now often rely on younger colleagues, who use the Internet and may not be aware of the older literature and services. We are going through a quiet revolution, though not, perhaps, the one that information scientists talk of.

Communication

Laboratory and clinical staff at the Imperial Cancer Research Fund, which has extensive information backup services, gave the highest priority in keeping abreast to talking, not the expensive database services:

Verbal	90%
Browsing	75%
Meetings	70%
In-house current awareness	52%
Fax, e-mail, written	40%
Preprints and refereeing	21% each
Online, electronic bulletin boards	5%
Other current awareness	4%

(From Davies, 1994.)

These staff may have better current awareness facilities than many, but the trend is clear: most scientists rely heavily on friends and colleagues, just as Balfour did. This is one reason why larger groups of scientists have an advantage. A large group will have seminars and outside speakers, more students and research fellows coming and going, forging links with other similar groups, more colleagues refereeing, speaking at conferences and generally socializing. Francis Crick suggested the gossip test: when scientists are at coffee, they gossip about their real interests in science. Meetings and seminars and journal clubs should be social occasions, with coffee and drinks, so that scientists mix and gossip. Conferences are necessary for meeting new and old friends. The lone worker or the small department should be helped, not dismissed.

The speed at which new techniques and ideas, especially in immunology and molecular biology, have been absorbed and put into use in the past two decades suggests that communication among life scientists has become faster, even if publication actually takes longer:

> In the February 1908 issue of Volume 10 of the *Journal of the Royal Army Medical Corps* there were two letters from overseas. That from Colonel J.G. MacNeece, dated 11th January, Malta, began 'Sir, – The January number of the *Journal* has just reached Malta.' The second letter, from Lt. Col. J. Barnett Wilson, dated 20th December, Alexandria, began 'Dear Sir, – I notice in the *Journal* of November last . . .'.
>
> Journals and mail were probably sent by steamer across the Channel, by rail through France to Marseilles and thence by steamer to Malta and Alexandria. There might be delays for suitable ships. Journals were typeset by hand and papers were corrected by authors from proofs. [Adapted from Wyatt, 1994.]

For specialist journals publication delays may now be as long as two years. For scientists in countries which have to rely on printed journals and secondary serials, there may be very long delays before these reach the library. However, even more damaging is our inability to use the information we have.

In 1995 Sir Donald Acheson spoke at a meeting at the Wellcome Institute about his work in medical relief in the former Yugoslavia, where refugee rations supplied by the world's charities are even now deficient in vitamins. The essential role played by vitamins has been known, and even before the First World War scurvy, beri-beri, pellagra and probably rickets were all thought to be caused by lack of vitamins. In the UK, the Ministry of Agriculture, Fisheries and Food and the Royal Society of Chemistry have a national nutrient databank, which is the equivalent of a complete update of the 1978 classic *Composition of Foods* by McCance and Widdowson. There is a databank with the composition of over 4000 food products for use by patients with food intolerance. Yet, at a 1991 conference on vitamin deficiency diseases in refugee camps, the prevalence rates for scurvy in Ethiopia were said to be among the highest this century. Rations in camps represented one-fifth of the lowest daily intake to prevent scurvy. In emergency medical kits for Kurd refugees in 1991 there was no vitamin A to prevent xerophthalmia (Toole, 1992). A subsequent letter showed that refugee rations were devoid of vitamins A and C, and low in niacin, iron and riboflavin (Tomkins and Henry, 1992): 'By contrast, pet food is not only balanced but also has excellent micronutrient composition.' Three years after these publications, in a journal which boasts half a million readers, our elaborate information systems give us healthy pets, but the information is not used by the international relief organizations.

Swimming not drowning

Editors, librarians and information scientists regularly deplore the increasing number of journals, but scientists have adapted and prospered. Attempts by some biochemists to bypass the refereeing process by using the Information Exchange Groups of the 1960s were unsuccessful because the ciculation of preprints actually increased the amount to read. Although D.E. Koshland's prediction of *One gene – one journal* in *Nature* 1983 has not come true, yet we can see the tendency towards single journals for animals, plants or microorganisms or for single diseases: a recent journal is for the single bacterium *Helicobacter* 1996–. Scientists have to specialize to conquer while skimming the rest of the literature for new, good ideas and techniques. Eugene Garfield and the *Journal*

Citation Reports list journals by impact factor (IF) – the higher the IF, the more important the journal. However, the important journals for a small specialist subject may have a low IF within a large subject grouping, for example biochemistry and molecular biology, with 151 journals (see p. 166). Recent attempts to increase research ratings through the use of bibliometric data such as IFs can lead to self-defeating edicts. Thus, one research institute publishes a respected specialist journal which has a low IF the managers are urging the staff to publish in journals with high IFs. As only the best papers, and possibly those least related to the aims of the institute, will be accepted by the journals with wider appeal, the managers risk their own journal and even their institute. As Sydney Brenner has said, 'citation often tells us more about sociology of science than about science itself' (Brenner, 1995). Prolific authors are listed by Anderson (1992).

In the same way senior Indian scientists and politicians, when considering applicants for promotion, only take notice of papers published in international journals. All except the very best Indian journals are thus condemned to publishing papers rejected by international journals, either because of quality or because the subject is of local interest only. Research is directed from India's needs to imitations of western science (see p. 101).

An interesting facet of citation analysis was discussed by Bernard Dixon (1994). Alison Rawlings has shown that, although scientists originally cited Gallo's paper rather than Montagnier's for the discovery of HIV, Montagnier's now receives more citations for the discovery. Scientists apparently vote not with their feet, but with their citations.

Nature and *Science* should be read every week. Although the contents pages are reproduced in *Current Contents* and can now be obtained via e-mail, there are sound reasons for suggesting that these vital sources of comments, news, important papers and new ideas should be looked at and the pages turned. However, all too often the main papers are difficult for experts, and even the titles mean little to the outsider, so that many scientists no longer look at *Nature*, except for the appointments pages. *Nature's Genome Directory*, a supplement of 400 pages, was distributed to institutional subscribers in October 1995. Looking at *Nature* and the other weeklies gives a good background to the topics of particular interest as the research frontiers move (see Table 1.1)

There are also *Nature Genetics*, *Nature Structural Biology* and the new *Nature Medicine* 1995–: this last will interest a few life scientists. *Nature* has product reviews, new journals reviews, a software review including statistics, as well as more traditional book reviews. Special issues appear in *Science* e.g. AIDS issue

Table 1.1. Change in distribution of articles about cancer in *Nature*

	% articles			
Topics	1975	1980	1985	1990
Biological and medical aspects of cancer	50	36	16	9
Cancer and viruses	30	48	35	19
Biochemistry and molecular markers of cancer cells	19	12	5	9
Oncogenes	1	4	43	64
N=	97	77	92	47

Adapted from Morange (1993). Figures for 1990 kindly supplied by Dr Morange.

May 1993 and Genome issue October 1993. *New Scientist* should also be read every week and from 1995 is available on CD-ROM, with each quarterly disc carrying 260 issues of the previous five years. Although expensive, this will make it much more accessible, particularly in libraries where the journal is frequently vandalized. However, the availability of CD-ROMs in libraries has greatly increased the use of retrieval tools and put severe pressure on their use. One may now have to reserve the use of a terminal or a particular service. *Nature* 1992–1994 is now on CD-ROM. Many health professionals said that *New Scientist* was their main source of information about HIV and AIDS (Wyatt, 1988).

As new ideas arrive words, acronyms and phrases to describe them appear, and it can be difficult to trace the exact meaning, or any meaning: *Biochemical Education* has a page of 'Lexicon – words and phrases at the growing edge'. Other journals, particularly the general ones such as *BioScience* and *Biologist*, should copy this feature.

Reviews

Although the commentary sections in the weeklies explain the recent papers, rather longer and more illustrated reviews appear in the monthlies, most of which are called *Trends in . . .* Perhaps surprisingly, these journals have some very high impact factor:

	Impact factor
Trends in Biochemical Sciences (TIBS)	13.5
Trends in Biotechnology (TIBTECH)	2.7
Trends in Cell Biology (TCB) [no category]	
Trends in Ecology and Evolution (TREE)	3.5
Trends in Food Science and Technology (TIFS)	–
Trends in Genetics (TIG)	10.0
Trends in Microbiology [too recent]	

Trends in Neurosciences (TINS)	17.3
Trends in Pharmacological Sciences (TIPS)	16.2
Immunology Today	19.6
Parasitology Today	3.7
BioEssays	5.0
Molecular Medicine Today [too recent]	

(From *Journal Citation Reports* 1993)

Although *Parasitology Today* has a low IF, this journal was top of the table for parasitology. Each journal now has annual author and subject indexes and *Immunology Today* is online from 1995; others will probably follow. *TIPS* has an annual Receptor and ion channel nomenclature supplement and a special drug index (1994). *TINS* produced a supplement on neurotoxins in April 1994. There are often excellent colour diagrams. *Microbiology Europe* (1993–) is an interesting free journal with short reports.

Scientific America is very useful because it provides understandable reviews with excellent illustrations and presentations of more general topics rather than immediate news. Several special issues have been reissued as books.

In the *Journal Citation Reports* the various annual review series have very high ranking in some of the important biological subjects (see p. 70): they are probably used as a general reference in place of a large number of individual ones. *Annual Reviews* is a well established series, with 30 or more reviews in each volume, of which one is usually an interesting autobiographical review. A diskette lists more than 4500 articles from 1986 onwards, as *Annual Reviews Index*: this is updated each quarter.

The gap between the short review of a topic in, for example, *Trends in . . .* and the comprehensive and scholarly review in, for example, *Annual Review of . . .*, which will have taken some time to complete and print, has been filled by several new series. *Current Opinion in . . .* is published six times a year, with each issue covering a single topic with perhaps 15 reviews of the recent literature. For each review, recent papers in the bibliography of references and recommended reading are starred for 'special interest' and 'outstanding interest', with short descriptive notes. Each issue contains a comprehensive bibliography, which is also available as an electronic database, BIOBASE, on diskette.

Current Opinion in Biotechnology 1990–
Current Opinion in Cell Biology 1989–
Current Opinion in Genetics and Development 1991–
Current Opinion in Immunology 1989–
Current Opinion in Neurobiology 1991–
Current Opinion in Structural Biology 1991–

All six titles also appear on a single CD-ROM. There are 24 similar serials in medicine, of which *Current Opinion in Infectious Diseases* and *Current Opinion in Endocrinology and Diabetes* will interest some biologists. However, *Current Biology* 1991–, from the same publishers, is a review journal with ordinary original papers.

Another new series has six issues per year, each devoted to one subject with a guest editor: there are fewer articles in each issue and the bibliographies take the usual format:

Seminars in Cell Biology
Seminars in Developmental Biology
Seminars in Immunology
Seminars in the Neurosciences
Seminars in Virology.

Another series is Baillière's International Practice and Research (BIPR) for medical scientists and clinicians: *Clinical Infectious Diseases* 1994– may be the most interesting for biologists.

The *Index to Scientific Reviews* 1974– is issued twice a year, with a four-month delay from the last entries. It is compiled from 225 review serials and from 3000 'leading' journals, with 30 000 reviews and review-type articles each year. The *Index to Scientific and Technical Proceedings (ISTP)*, on CD-ROM in the first issue, has 800 000 papers from 18 000 conference proceedings over the previous four years. Quarterly updates each contain about 45 000 papers from 1100 recently published conference proceedings. Both indexes are based on the SC1 database (see p. 78).

Digests of the news

A recent trend has been the need of some scientists, managers, politicians and others, for digests of the news in particular subjects, for example AIDS, the environment and biotechnology. The weeklies *BMJ, Nature* and *Science*, for example, now have notes highlighting the most newsworthy papers and items so that science correspondents can easily see the week's stories. Scientists will find their specialist interests through *Current Contents* and perusal of journals. Health workers and managers, activists, politicians and others concerned with policies for AIDS must be aware of the latest news, but are unlikely to see, let alone read, the masses of literature in the medical and scientific publications. They need a digest of the significant new findings which includes social and political views. Between 1985 and 1988 at least six newsletters were founded in the USA, one in Canada and six in the UK, as well as the AIDS Monitor section in the *New Scientist*. They ranged from weekly to quarterly and from 1200 pages a year to only 30.

The staff of the Division of Human Health, a part of CABI, produce *AIDS Newsletter* and *Current AIDS Literature*, both monthly. The former provides 'an authoritative, accurate and up-to-date synopsis of global developments on HIV/AIDS as reported by the media and the medical and scientific press'. Each issue has an editorial, news, guest articles and conference reports, social and occupational aspects etc. The latter has abstracts of papers, books, reports et., with editorial comments from a panel of experts. The standards of these publications are very high (see p. 84). The monthly *Public Health News* is also printed in *Abstracts on Hygiene and Communicable Diseases*. Items are culled from newspapers and journals, with a delay of about one month. Subjects include food and nutrition, the spongiform encephalopathies, disasters and water quality, as well as more medical aspects.

Biotechnology is a subject with great financial potential, so that all kinds of managers want news and informed comment. There are many digest newsletters (see p. 133), which provide good reading and appraisals for scientists and non-scientists. *GenEthics News* has a wider public audience (see p. 141). Whether there is a market for more scientific audiences has not really been tested. Perhaps subscription services on the Internet might be cheap enough to sustain science digest newsletters.

Meetings

Notices of meetings may be found in the monthlies and the product magazines, as well as in many society journals, such as *SGM Quarterly*. EVENTLINE is an online database, part of EMBASE (see p. 82), of all kinds of meetings, trade fairs etc. For biotechnology and related interests see p. 134. Unfortunately, many of the notices appear so soon before the event that it is difficult to see how people would be able to attend unless they lived in the same town and could easily rearrange their other commitments. Greater notice would be very welcome.

References

Anderson C. Writer's cramp. *Nature* 1992; 355: 101.
Brenner S. Loose ends. *Current Biology* 1995; 5: 568.
Davies M. The information handling techniques of research laboratory and clinical

scientists: preliminary indications of current practice. *Health Libraries Review* 1994; 11: 183–195.

Dixon B. Commentary. *Bio/Technology* 1994; 12: 1050.

Morange M. The discovery of cellular oncogenes. *History and Philosophy of the Life Sciences* 1993; 15: 45–58.

Tomkins A. and Henry C.J.K. Comparison of nutrient composition of refugee rations and pet foods. *Lancet* 1992; 340: 367–368.

Toole M.J. Micronutrient deficiencies in refugees. *Lancet* 1992; 339: 1214–1215.

Wyatt H.V. *AIDS Information: Supply and Demand*. British Library Research Paper 55, 1988.

Wyatt H.V. Publication delays in the age of steam. *Lancet* 1994; 343: 416.

CHAPTER TWO

Contents pages of journals

H.V. WYATT

Current Contents (CC), in the two versions *Life Sciences* and *Agriculture, Biology and Environmental Sciences*, is based on about 1350 and 950 journals, respectively. New journals are not automatically added, but have to submit copies for evaluation. As new journals are added others are dropped, so that as the number of journals increases the proportion covered decreases. The weekly printed issues of *CC* are sent by air freight and may arrive several months before the journals themselves. The indexes are taken from the words in the paper titles. There is a *Quarterly Index to Current Contents/Life Sciences. Current Contents on Diskette* has an index which includes words and phrases from the bibliography. Another version has author abstracts for about 90% of the articles. The different versions are also available on CD-ROM, with a weekly rolling file for the previous year and an annual archival disk. The *Scisearch* database may be searched online.

Reference Update 1988– for biology and medicine is available online, via the Internet and in a variety of disk formats. The weekly Basic edition covers more than 400 journals; the weekly deluxe edition and the Deluxe edition with abstracts cover more than 1200 journals. There is a monthly Clinical edition, covering more than 100 journals, which is not available on the Internet. This service offers integrated use of several sophisticated software programs, e.g. Reference Manager and Grant Manager (see also p. 89).

Inside Information on CD-ROM 1993–89 contains details of articles from the contents pages of 10 000 of the most frequently requested journals from British Library Document Supply Centre (BLDSC), with 1 million articles a year (see p. 98). It is updated monthly, with annual cumulations. The contents pages are available two to six weeks from receipt of the journal. This is an excellent current awareness tool covering all the sciences, not just the life sciences. The BLDSC also offers a journal contents page service, of journals selected by the buyer from the 76 000 serials currently received, which are listed in *Current Serials Received* (see p. 98). For India see p. 102.

Medexpres 1986– has the contents pages of French-language journals, including those from French-speaking Africa, e.g. Togo and Tunisia. There is a subject index.

SwetScan is a new major electronic current awareness service based on 12 000 titles in all subjects, available weekly as printout, diskette or tape, or continuously through DataSwets or via the Internet. The service will soon include abstracts. There are about 1000 biological journals and the complete *Zoological Record*, with other journals of interest listed in the medical sections. In a printed version only, serials are listed by subject and title.

SCAN: Life Sciences and Medicine is a listing of the contents of the journals published by Gordon and Breach and Harwood Academic, with about 30 titles of interest to biologists. The range of interests is very wide. *SCAN Plus: Life Sciences and Medicine* 1994– gives the abstracts of papers from some of the journals, but there are no page numbers and although both serials are described as 'current awareness' there is a long time lag.

Most of the services offer a reprint service, either as tearsheets from the journal or as photocopies. UnCover is a similar service covering 15 000 journals from Blackwell's in Oxford.

Searching some form of contents pages may be a very good way of keeping up with the literature. However, one must check which journals are covered. One may look for the contents of particular journals of known interest, and it would be best to use the subject index as well. The subject index is based on words in the title, whereas many secondary serials use not only the abstract and any given keywords, but also a scan of the contents of the paper itself. Many journals, unfortunately, do not list the titles of brief notes and letters on their contents pages. Other journals have announcements of meetings, grants, bibliographies, 'interesting snippets' etc. as space-fillers within the body of the journal. There is no substitute for browsing through the journal itself.

Selective services

Most of the services also offer database searches tailored to the needs of individuals or groups. Others offer bibliographies on particular subjects. SUBIS, now part of Sheffield Academic Press, sells Bulletins produced by manual scanning of 2500 journals by scientists. There are about 100 different subjects in the life sciences and biomedicine, with particular emphasis on biotechnology, cell biology, immunobiology, neurobiology and physiology. One advantage of such a broad range of subjects is that bulletins, because they are simple with no abstracts, can be added or deleted to match new subjects of interest. Titles range from 'Activin and inhibin' to 'Tissue culture', 'Plant biotechnology', 'Immunoassay (with annotations)', '*Helicobacter*' and 'Apoptosis'.

The US Department of Commerce includes the National Technical Information Service (NTIS), which produces some 200 *NTIS Alerts* every two weeks on R & D science subjects, adding 70 000 items each year from 100 countries. The sections of interest are agriculture and food, environmental pollution and control, including pesticides, medicine and biology, biological oceanography, forestry and soil sciences. There is also a section on chemical, biological and radiological warfare. About one-tenth of the sections are relevant to the life sciences.

CHAPTER THREE

Electronic information sources and the future

H.V. WYATT

Medical journals, especially the weeklies, have large readerships and advertising revenues and can fund staff for investigations and publishing projects. The editors of some of the most prosperous journals formed the Vancouver Group, which has issued many guidelines that are accepted by several hundred journals. Unfortunately, few of the non-clinical and more biological journals are so far involved with this group (see also p. 89). A standard format for manuscript submission has been followed by statements about responsibility, retractions, fraud, ethics etc. Journals such as *The Lancet* have offices in the USA, with simultaneous publication in two continents; there is also a French edition. The *British Medical Journal* has student and Indian editions. *Pediatrics*, an American journal, has foreign-language editions in Spanish, Italian (1990–) and Portuguese (1993–). Since 1992 the Italian edition has included *Pediatrics in Review (PIR)* and commentaries by Italians. *PIR* is published in Spanish in Argentina. *Pediatrics* is published in India in English (1990–); there is an Arab world edition 1992– with a selection of articles and abstracts of others in Arabic, and is expected to have commentaries in Arabic. *New Scientist* is published simultaneously in Britain and Australia. It is difficult to think of life science journals which would have sufficient appeal for any similar ventures, although journals such as *Biochemical Education* and *Journal of Biological Education* could arrange for relevant papers to be printed locally in places such as India, China and South America.

Although the larger academic societies, e.g. the Society for General Microbiology (SGM) in the UK and the American Society for Microbiology (ASM), have long published symposia and monographs as well as journals, more and more societies and other institutions are publishing books. Cold Spring Harbor has become a major publisher, as has the old Commonwealth Agricultural Bureau (CAB), which is now known as the Centre for Agriculture and Biosciences International (CAB International). As well as the abstracting journals CAB International has a list of over 300 books, including directories, reviews, keys, databanks, dictionaries and series of papers. The British Crop Protection Council and the Royal Society of Chemistry now jointly publish the *Pesticide Manual*, the *European Directory of Agrochemical Products (EDAP)* and *The Pesticide Index*, and plan new computerized publications.

The Oxford Medical Databases will interest clinical geneticists and neurologists, and could be the model for similar schemes for biologists. More than 1000 journals are reviewed for *The London Dysmorphology Database* and the *London Neurogenetics Database*. These may be searched by menu or keyword, and together contain information on nearly 5000 syndromes, with abstracts and approaching 40 000 references. There is an associated *Dysmorphology Photo Library* on compact disc. Comparable databases might become tools for systematic and developmental studies, or for groups such as enzymes.

What's new with proteins

The subject of proteins illustrates the wide range of different publishing formats which have recently appeared. No doubt other disciplines will produce similar diversity. The journal *Protein Sequences and Data Analysis* publishes original papers and protein databank entries. *Human Protein Data* edited by A. Haeberli, (VCH 1992–) is a series of annual looseleaf compendia with a computer retrieval program on diskette, with 300 proteins by 1994. *Protein Profile* 1994–, with at least 15 issues a year, will cover a specific protein family in each issue, with an updated version each year. Each issue will include a comprehensive bibliography and special fold-out pages for viewing sequence alignments. Each issue of *Protein Science* 1992– includes a floppy disk with abstracts and references for all papers, an interactive graphics supplement, a graphics program and the 'PREKIN authoring tool'. The CD-ROM version includes the Brookhaven Protein Data Bank and other databases. *Macromolecular Structures* 1991– is an annual which features every structure determined by X-ray crystallography and/or nuclear magnetic resource: it is available on CD-ROM with text tables and images, with additional kinemages of many structures. SCOP is a World Wide Web (WWW) database on the structural classification of proteins. *Molecular Structures in Biology* is a CD-ROM of the more than 500 structures of the Protein Data Bank, with images of them.

RNA 1995– can be reached on the World Wide Web for contents and forthcoming papers, as well as editorial notes.

Electronic publishing

Electronic distribution of data already occurs and preprints of articles are available. However, not all scientists have access to the networks, and charges may come. Publishers of journals obtain advertising revenue, which for some journals may be very considerable. The experimental Information Exchange Groups in biochemistry in the 1960s foundered because of the vast bulk of the preprints, and because changes occurred in the refereeing and editing processes: titles as well as contents were changed, mistakes were rectified. It was difficult to know what had been published, and many were outside the organization. Electronic distribution could lead to a salami explosion of unrefereed and dubious work. Publishers might be happier with printing, for a fee, at library terminals, but this might leave individual subscribers in difficulty.

Human Reproduction Update 1995– began with six issues planned per year. Each printed issue comes with a CD-ROM disc containing graphics, videos, databases, 'new terms defined' and a cumulating bibliography. The journal incorporates the reviews begun in *Human Reproduction*, the *Bibliography of Reproduction* and the *Oxford Reviews of Reproductive Biology*. The *British National Formulary (BNF)*, an essential reference for every doctor, is now available on disk for use with a personal computer. CD-ROM provides the means by which several products can be combined for the benefit of the user, e.g. *Methods in Enzymology* (p. 41).

The Internet, CD-ROM and journals

The BIOMEDNET is a WorldWide Web club for biomedical scientists, launched in 1995. For an annual fee users will be able to download journal articles from publishers, who will make charges in line with photocopying. There will be facilities for 'meeting rooms', virtual lectures, noticeboards etc. and a 'shopping mall' with catalogues etc. for buying products and services. The organizers are the publishers of the recent series *Current Opinion in . . .*, all of which are available to users (see p. 6) and are available as a single CD-ROM updated three times a year. See also p.165.

An ambitious project is Bioline Publications, which uses an office at Bourne, Lincolnshire, UK, and a computer at the Tropical Data Base in Brazil. The services are available via e-mail, WWW and gopher. All tables of contents, abstracts and summaries are free, as are several journals; payment is required for delivery of the full text and graphics. Other journals are by subscription. There are initially about 30 journals, many EC and US documents, reports, information about culture collections and 20 plus newsletters, of which several are research newsletters, such as *Polychaete Newsletter*. So far the emphasis is on biotechnology and biodiversity, but clearly the range will expand as publishers and groups add products. There are opportunities for small publishers, groups and authors to use BIOLINE to reach a world audience. Among those already using BIOLINE are the African Centre for Technology Studies, CAB International with their *AgBiotech News and Reviews*, International Council of Science Unions (ICSU), Oswaldo Cruz Institute, University of Amsterdam and White Horse Press with *Environmental Values. BioSafety* 1995– *Biopolicy and Sustainable Development* 1996– and *Biofilm* 1996– are peer-reviewed journals only available online, with others planned. Three books are available and others will follow. *Bioline* has the potential for making good material from developing countries available without the usual difficulties of selling to libraries and mailing costs and delays.

A number of databases, e.g. the Life Sciences Collection (p. 78), may be accessed via the Internet through annual subscriptions to SilverPlatter.

A computer conferencing network using e-mail has been used by a microbiology group, linking researchers in South America and Canada with others worldwide. The exchange of information has mainly been about techniques of reagent production and purification: see *Networking in Brucellosis Research*, edited by J.F. Frank, (Tokyo, UNUP, 1991).

Florida Entomologist is now online on the Internet at no cost to the user, each article being a Portable Document Format (PDF) file. The cost to the parent society is less than $5 a page, which can be recovered as a page charge to the author. Alternatively it can be introduced as a charge for the full journal, or for individual papers as an electronic reprint service. The *BMJ* had a series of articles on the ABC of Medical Computing in Volume 311, 1995, which will probably be published under that title: a useful article on the Internet appeared on pages 440–443. Microbiologists will wish to consult the two articles in *ASM News* 1995; 61:

Knight P. Microbiology resources on the Internet. 397–400.
Terry TM. Teaching microbiology with the World Wide Web. 401–405.

The *E. coli* index is a series of pages on the WWW with details of microbiology resources. There is access to data on *E. coli* genes and proteins (see p. 145) and pathogenic strains:

http://sunl.bham.ac.uk/bcm4ght6/res.html

Although in theory the Internet is available anywhere, there are still substantial limitations in many parts of the world. In India the Internet was limited by the government to universities and research institutes, with only about 30 000 users, but in 1996, access was liberalised with an explosion of users.

The *Directory of Electronic Publication* 3rd edn, Association of Research Libraries, USA, has details of 240 electronic journals, newsletters etc. and 1142 scholarly lists:

Ann Okerson ann@cni.org

10 Minute Guide to the Internet, P. Kent (Indianapolis, Alpha Books, 1994). For computer-aided learning see the special section in *Biochemical Education* 1994; 22: 40–44. *The 1995 Internet White Pages*, S. Godin and J.S. McBride, is an annual from IDG Books, with three updates per year, listing about 100 000 Internet addresses alphabetically by name and domain.

One difficulty is that addresses often change, and navigating the Web is daunting for newcomers. Fortunately, many librarians and information scientists are preparing directories and guides for the subjects of likely interest to the staff of their institutes. Society newsletters and journals will be good sources of information, for example the newsletter of the Royal Society of Tropical Medicine and Hygiene for parasitology, and *Biologist* 1996; 43(1): 12, which listed Uniform Resource Locators (URLs) for some biological WWW sites. Conferences will have facilities and demonstrations. The Information Centre of the Wellcome Centre for Medical Science in London is an important resource. Articles in the *BMJ* ended with a guide to the Internet and how to access an electronic version with additional hypertext references (Pallen, 1995). In general, the medical facilities with more money are more advanced than those in pure biology.

Some journals are available on CD-ROM, but most of these

are medical journals with very large circulations, even more readers and large resources, for example *The Lancet*, *New England Journal of Medicine* and *Infectious Diseases*. Few life science journals are yet on CD-ROM: *Development* may appear in this form in 1996, but in certain fields there are exciting packages. The journal *Virology* appears quarterly on CD-ROM, with an annual disc replacing the others. The *European Journal of Biochemistry* 1993– is updated semiannually: as with some other items there is a very low subscription if taken with the printed version. The savings in binding, storing, shelving and library space may make this an attractive proposition for libraries, provided there are enough CD-ROM readers for use throughout the day or when needed. The 11 journals of the American Society for Microbiology (ASM) are available as annual archival discs 1992– and as ASCII texts from 1989–. The *Journal of Biological Chemistry (JBC)* is now on CD-ROM quarterly, with full colour graphics. It is possible that it will only be published online from 1998, with no printed version.

As in many other spheres there are enough potential buyers in medicine to permit expensive projects; whether there would be enough buyers in the life sciences for similar initiatives remains to be seen. The *Multimedia Encyclopedia of Mammalian Biology* is a CD-ROM with 3500 colour images, 500 maps, sound and text. It is based on the five volumes of *Grzimek's Encyclopaedia of Mammals* (McGraw-Hill, 1990) and the BBC Natural History Archives with considerable additions: it appeared in an updated second edition in 1995 with newly designed software. The *Oxford Textbook of Medicine*, edited by D.J. Weatherall, J.G.G. Ledingham and D.A. Warrell (3rd edn 3 vols 1995) is available in an electronic edition on compact disc, but without tables and figures although enhanced captions are included. It can be updated more easily in this form than in the printed version; occasional items are reprinted in the *BMJ*. Five electronic indexes give access to all the material. However, one library which has both the printed and CD-ROM versions has found that the books are much preferred and the CD-ROM version is neglected: the latter is difficult to browse through and use of the index produces long lists of pages. Oxford Electronic Publishing, part of OUP, also produces the *Oxford Database of Perinatal Trials* and *Leucocyte Typing Database V*. HyperCELL by G. Williams, Garland, is an electronic companion to *Molecular Biology of the Cell* by B. Alberts *et al.* It is updated annually, together with references for use with Windows and Mac. No doubt other popular teaching texts will follow.

The large medical publishers, Raven, Lippincott, Churchill Livingstone and Mosby, have many books on CD-ROM. *Mendelian*

Inheritance in Man: a catalogue of Human Genes and Genetic Traits, by V. McKusick, is now in an electronic version MIM-CD for use with a PC or Mac (see p. 143). Searching is with fuzzy-logic software. MEDLINE discs 1966– are also available for use with the same software. Journals with full images are also promised from Lanmark Medical Innovations, Beaconsfield, UK, or Aries, Andover, MA, USA.

Many journals now have their contents pages and other information online, for example all Chapman & Hall journals have contents and abstracts free online from 1996, as do *Journal of Molecular Biology*, *Nature* and *Science* etc. The advantage for the specialist journals is that their contents are available online about four weeks before the printed copy. See also p. 165.

Reference

Pallen M. Introducing the Internet. *British Medical Journal* 1995; 311: 1422–1424. See the box on p. 1423.

UK Guides

Two short guides are available from The Wellcome Trust Information Service, 181 Euston Road, London NW1 2BE:

Finding Information on the Internet

The Netscape Navigator Web Browser, which has a short glossary of Internet terms.

Both guides were produced in October 1996 and will probably be updated when necessary.

JANET 1995 a library overview is a short guide to the UK Joint Academic Network with details of, for example, BIDS (see p. 69), library catalogues, university etc. information including e-mail addresses of staff.

Internet Resources for Medicine and Bioscience is produced by OMNI, Organising Medical Networked Information based at the National Institute for Medical Research London NW7 1AA, and at http://omni.ac.uk/ Although the emphasis is, of course medical, there are addresses for viral names, culture collections, Online Mendelian Inheritance in Man, *BIOSIS Previews*, BioSupplyNet, *Journal of Biological Chemistry* table of contents and many others of interest to some biologists.

The Student's Guide to the Internet, by I. Winship and A. McNab (London, Library Association, 1996) is very general with little for the life scientist as such, but has eight examples of how to cite electronic sources and a guide to creating a Web page.

CHAPTER FOUR

New ideas and fraud

H.V. WYATT

Where do new ideas come from? I wish I knew. They come from rearrangements prompted by some new stimulus. They come through recognition of patterns, of parallels, analogies and variations based on past experience and new information. The more limited our experience, the more limited our store. If we limit our reading to just our present experiments, we limit the chance for new ideas. It is no accident that many academics are good at teaching and research, in spite of the time devoted to teaching. Teaching, explaining a wide variety of phenomena, provides new experiences, more rearrangements, the opportunity to see the parallels and to be questioned and questioning. There is a danger that the scientists sitting in their laboratories and searching online will have no incentive to walk to the library. They will miss the books. They will miss Minerva on the last page of the *BMJ*, with her clips from the journals. People who change disciplines bring not only new techniques but also new patterns, new ways of thinking – but they have to know enough already to want to change and where to change to.

Serendipity is all around us, but we have to build on a broad base of experience. Some information scientists have suggested that the way to deal with the increase in the literature is to limit each scientist to one or two papers a year. This is nonsense. We have to learn to write, just as we have to learn to experiment. If we cannot publish our work, why do the experiments?

An older generation of scientists had hobbies which often involved an interest in some form of natural history. Hobbies often provided the parallels or the skills for the experiments.

Joshua Lederberg (1989) has written that ‘Today, even more

than ever, many more ideas are afloat than any mortal can reasonably attend to. [The ideas] most vulnerable to deterrence are those that require a spanning of disciplines beyond customary teaching.' I would add 'or customary reading'. It has been suggested that some ideas, like those of Avery and of Garrod, are not accepted because they cannot be fitted into the accepted paradigms. I have suggested (Wyatt, 1972, 1975) that in the case of Avery, his findings were not immediately assimilated because there were no obvious follow-on experiments, and the same applies to Garrod. Lederberg (1989) has pointed out that no-one realized that sickle cell disease was a parallel to Garrod's inborn factors of disease.

Fraud

Instances of fraud and misconduct are occasionally reported in *Nature*, but fraud has not been seen to be as big a problem in the life sciences. In medicine and the related sciences there has been much concern, with editorials and articles in, for example, the *British Medical Journal* and *Fraud and Misconduct in Medical Research*, edited by S. Lock and F. Wells (BMJ, 1993; updated edition 1995). In the USA there have been enquiries by Congress and the National Institutes of Health (NIH), and the problem is taken very seriously. Several articles on the 'conduct of science' were published in *Science* 1995; 268: 1705–1718 with plans for a continuing discussion online on the WWW. *Scientific Integrity*, by F.L. Macrina (ASM 1995) includes the use of animals, scientific record keeping etc. and should be a good reference book for every doctoral student.

Once papers are accepted and enter the journal literature, there is great difficulty in tracing errors which may be corrected or frauds which have been exposed. As yet only *Index Medicus* and its associated services flag the original paper with such corrections. Other life science serials do not.

The pressures for advancement in clinical sciences have been very great and it may be difficult to repeat clinical observations before the author has moved on. In the life sciences the pressures may be as great, but supervision has usually been more thorough: this may change with new financial pressures. More important, perhaps, are wrong observations and analyses, bias, self-delusion, manipulating data and suppressing inconvenient facts, observations or experiments. We should all read articles very carefully and critically, as piracy, plagarism and forgery do occur. Referees have a special position of trust and editors should ensure that referees return papers very promptly, and that reasons for rejection are valid. The Internet has a discussion group on fraud in science, SCIFRAUD.

Doctoral supervisors often co-author papers – it is to hoped as the second author – because the work has originated with them and they have made significant contributions. It is, however, a common practice for the head of a department to insist on being named an author on every paper from the laboratory. This leads to others wanting to be co-authors as well. Some journals require a letter detailing the contribution of each author and signed by all. Some require no more than four authors. One journal requires a signed letter from a head of department attesting to creative participation before listing a person as a co-author. Gift authorship, whether given by a real writer or accepted under duress, is bad for science. There have been several cases where senior and respected scientists have been held responsible for fraud or misconduct because they had been named as co-authors. This is a matter which should be debated in life science journals. We must ensure that due credit for supervision and fund raising is acknowledged – and recognized – without authorship. This requires a change in the criteria used by funding bodies and by institutions when considering promotions.

In the five years 1983–1987 there were 8926 AIDS-related articles in the MEDLINE file. A single person was the co-author of 177 of these (single-authored papers were not included). This represents 35 papers per year or one paper every five working days, assuming holidays and attendance at meetings (see p. 4).

It was common practice for microbiologists to gather dust when visiting commercial works, for later cultivation. Now, outright theft of genetically altered organisms could be possible.

Many journals publish only original papers, sometimes augmented with short reports, but few publish correspondence about such research. Journals such as *The Lancet* and the *British Medical Journal* have extensive, widely read correspondence columns, although the *BMJ* received about 3800 letters in 1996 and published 1500, from a circulation of 110 000 and an even larger readership. Even so, few letters are comments on original reports. Although original articles are widely cited, citation of correspondence arising is rare: letters in a later issue of the journal are rarely found by searches because indexing serials, except for *Index Medicus*, do not link correspondence with an original entry. A good example was a letter criticizing the design, data, interpretation and conclusions of a paper on snoring and disease. *Science Citation Index* lists 70 citations for the paper and one for the letter. The *British Medical Journal* letters are indexed and linked to the original paper in both subject and author indexes. One suggestion is that every published paper and editorial etc. should have a unique number, which would also be used for any letter, comment, correction or retraction (Bhopal and Tonks, 1994). Such a unique codon could be used by indexing serials, for ordering reprints and by editors in tracking manuscripts.

All journals should have a correspondence section, with the possibility that the authors of the original papers have a right of reply.

References

Bhopal B.S and Tonks A. The role of letters in reviewing research. *BMJ* 1994; 308: 1582–1583.

Lederberg J. Preface. *In Garrod's Inborn Factors in Disease*, ed. CR Scriver. Clarendon Press, 1989.

Wyatt H.V. When does information become knowledge? *Nature* 1972; 235: 86–89.

Wyatt H.V. Knowledge and prematurity; the journey from transformation to DNA. *Perspectives in Biology and Medicine* 1975; 18: 145–156.

CHAPTER FIVE

Newsletters and invisible colleges

H.V. WYATT

Many irregular and regular serials are called newsletters. Some from manufacturers are promotional literature; some from societies and groups carry news and ephemera. What I have called research newsletters are very different and are the invisible college in print. In general they are not publications, as they only circulate within a group, are often not available in libraries, except departmental ones, and are not listed as serials. I distinguish them on the basis that they contain one or more of the following unique items: a bibliography of publications by the college; short notes not published elsewhere; lists of strains and mutants etc., with location and nomenclature. They often appear irregularly and are usually associated with a particular organism or group.

The bibliographies are particularly useful as they represent almost complete listings of all the publications of the group, whether from journals, books, theses, conference proceedings or obscure foreign journals. Because members send in their own publications as soon as they are accepted, these bibliographies are more timely and complete than those of secondary serials, and would be very difficult to put together even if all secondary sources were searched.

The first research newsletters were *Maize Genetics Cooperation News Letter* 1932– and *Drosophila Information Service (DIS)* 1934–. These newsletters arose because researchers needed access to mutants and other stocks, and to formalize the naming and location of the products of mapping. Eventually, the groups became large enough to have meetings and to fund stock centres, and for data to be published. Now, databases of the information are available (see p. 144). In 1986 there were more than 110 research newsletters being published (Wyatt, 1986, 1987). Given the importance of sequencing of nucleotides and polypeptides, it is necessary that mutants etc. are available to other research workers.

For the first two decades almost all the research newsletters were based on genetics, but since then more and more are based on specific groups of animals and plants; the international *Perla* and the North American *Rolling Stonefly News* cooperate in meetings with each other and a group researching mayflies. Research newsletters often have obituaries of prominent members, with full bibliographies. An important resource for zoologists and botanists is the collection of specimens developed by individuals or brought together in museums or departments. Such collections may be difficult to find or to rescue on the death of the collector or with changes in a university. The research newsletter is one way of keeping track of such collections. One important function of the research newsletter is to publish short findings and notes on methods etc., which would not be publishable in the normal peer-reviewed literature but would be useful for other researchers.

Barbara McClintock received little recognition of her findings in research journals on jumping genes, and so from 1943 to 1977 she published 30 of her 39 papers in the Year Book of the Carnegie Institution of Washington and the *Maize Genetics Cooperation Newsletter*.

Despite the common belief that newsletters mutate into journals, I know of only two which have done so, *Arabidopsis* and *Mouse Newsletter*.

Research newsletters are a flexible and economical form of communication arising from the needs of geneticists to exchange mutants and strains and to agree nomenclature. When hybridomas were first made there were similar needs, and a newsletter was founded. However, those with hybridomas were asked for them by those without. As this caused an imbalance between supply and demand, which gradually widened, the newsletter was discontinued. *Fingerprint News* 1989–1993 was 'set up as a cheap, cheerful, fly-by-night operation, and like any good *r*-selected organism, when it has exhausted its temporary resource patch it is gone'; it was also the only research newsletter I have found with paid advertisements. It had a circulation of more than 400 and the last issue was a bibliography of 1257 references. The editors were approached by publishers, but preferred to allow other journals to flourish.

Ancient DNA Newsletter 1993– has a surprisingly large circulation of 700. There has been a complete archive bibliography 1970– and there is a more extensive listing through e-mail. There is an exceptionally wide range of journals with articles of interest, including Chinese and Japanese. The newsletter contains protocols, research news, articles and editorials, news of meetings and personals.

Producing research newsletters has been greatly simplified by photocopying, disks, e-mail and desktop publishing. Now, the Internet may make them obsolete. However, a survey of users of *Fingerprint News* showed that they preferred the printed version which they could read anywhere at leisure and file.

Small groups interested in a particular animal or plant come together socially to exchange information, organisms, rearing tips and much else. A group may produce a social newsletter, a research newsletter, a small journal of limited circulation, or combinations of these. Phasmids, or stick and leaf insects, are the core of one such group. The Phasmid Study Group in the UK produces a *Newsletter* 1980– of about 50 pages each year, which contains social and research news for about 450 members. *Phasmid Studies* 1992– is a small journal with an ISSN which is circulated to members and contains research papers and abstracts of the literature (see p. 221). Livestock is exchanged and distributed through the group. There are research newsletters produced by Dutch and French groups working on Phasmids. Groups such as these combine academics, professionals and amateurs, often have members from many countries, and publish studies from all over the world. See also p. 105.

References

Wyatt H.V. *Research Newsletters in the Life Sciences: their Scope, Advantages and Future.* British Library R & D Report 5897, 1986, 104.

Wyatt H.V. The invisible made visible. *Nature* 1987; 329: 357–358.

CHAPTER SIX

Finding books, journals and people

H.V. WYATT

Books and journals are sometimes difficult to locate, especially if only the journal abbreviation is given, for example *Me Nat*, which is actually *Maine Naturalist*: this may obvious to an American, but is not to a European.

Books

Bowker/Whitaker Global Books in Print Plus is a CD-ROM of over 2 million books in print from 90 000 publishers in 60 countries. *Boston Spa Books on CD-ROM* lists 700 000 titles collected by BLDSC since 1980. On microfiche this is called *Books at Boston Spa* (see p. 98). See also p. 178.

Journals

A good start is to look at the lists of journals to be found in the secondary serials, e.g. *Index Medicus*, where there are two indexes of journal titles, full and abbreviated. If this is unsuccessful, go to *Ulrich's International Periodicals Directory* 33rd edition 1994–95, available in print, online, on tape, microfiche and CD-ROM. Among the details are indexes for serials on CD-ROM and online, as well as editorial descriptions of many titles. Ceased and changed titles may be traced.

Boston Spa Serials on CD-ROM lists 500 000 serial titles and holdings at BLDSC and at other major British libraries. *Current British Journals* has over 8300 serials in UDC order. The latest edition has a ceased publication index, an index to sponsoring organizations and an abstracting serial index. *Current Serials Received* lists the 76 000 serial titles, including Cyrillic titles and cover-to-cover translations from the Cyrillic script.

Conferences

Boston Spa Conferences on CD-ROM lists 335 000 proceedings 1787–. *Inside Conferences on CD-ROM* 1994– provides details of 500 000 papers from the 15 000 conference proceedings collected each year. It is updated quarterly, with an annual cumulation. The *Index of Conference Proceedings 1964–1988* contains 250 000 conferences and is on microfiche. One is often confronted by an acronym and little clue as to its meaning: *Directory of Acronyms* (3rd edn, BLDSC, 1997) lists nearly 10 000 acronyms from the Conference Index Database of more than 350 000 records. About 30% are non-standard and very unlikely to be quoted in any other publication. However, the acronyms will be of societies, groups etc. which have sponsored conferences.

All the BLDSC databases may be searched with a variety of tools, and comments etc. added and printed or saved. They are also available in print, on microfiche and online.

Conference Papers Index 1972– from Cambridge Scientific Abstracts (p. 78) covers most of the sciences except physics, adding about 72 000 citations per year. A conference locator lists all the papers given at a conference and the contents lists information by field. There are subject and author indexes. There is an annual cumulative index, which may be ordered separately. The product is available in print, on magnetic tape, online and on the Internet WWW.

Other literature

SIGLE is a CD-ROM of the *System for Information on Grey Literature in Europe* 1980–, which contains reports, dissertations and other grey literature not usually accessed by the science secondary serials. The CD-ROM has 400 000 items, adding about 40 000 annually, and is updated semiannually. About half the entries are from the UK and Eire, but it is likely that more items from other European Union countries will be added in future. Unfortunately, only about 15% of the items cover life sciences and medicine. The British Library has rationalized its collections, which will now be concentrated at the BLDSC at Boston Spa. *British Reports,*

Translations and Theses (BRTT) will contain details of acquisitions (see p. 98). All items will be catalogued and available on the Online Public Access Catalogue (OPAC), both in the UK and eventually worldwide. *Information Sources in Grey Literature* by C.P. Auger (3rd edn., Bowker-Saur, 1994) is a recent guide, as is *Information Sources in Official Publications* by V. Nurcombe (Bowker-Saur, 1996).

Theses

Every year tens of thousands of master's and doctoral students write theses and dissertations, which fill the shelves of libraries and departments. Some are of poor quality and many have a limited interest, as the literature surveys and even the experimental results will soon be outdated. In the competitive experimental sciences theses are probably of the least value: the results will have been published in journals as soon as the work was completed and before the thesis was written. Nevertheless, some theses, particularly in ecological studies, for instance, contain valuable information which may be difficult to find elsewhere. Some research newsletters contain information on theses, and members of an invisible college will be aware of students finishing theses from presentations at meetings. *Dissertation Abstracts International (DAI)* Section B, covering science and engineering, every month provides abstracts of theses from participating universities in North America, in print and on microfiche. A quarterly international section, in print only, is published. There are author abstracts for doctoral theses 1980– and for master's theses 1988–.

Dissertation Abstracts Ondisc (DAO) is a series of CD-ROMs; the current edition, Section B 1987–, is updated quarterly. There are archival idexes going back to 1861 which were of North American institutions until quite recently, when some in Europe began to be included.

The BLDSC at Boston Spa, UK, holds 110 000 UK and 450 000 US doctoral theses (see p. 100). Details of British doctoral theses held on microfilm are recorded in *British Reports Theses and Translations* and online in the SIGLE database. One serious difficulty is that when theses are cited in the literature the DAI abstract number is not given. All citations to theses should contain not only the DAI abstract number, but also the faculty and department.

DOC-THESES is a CD-ROM of more than 300 000 French theses since 1972, in all disciplines. Health sciences, were added 1983–. There is a *Catalogue des Thèses de Doctorat Soutenues devant les*

Universités Françaises 1884–. *Sciences biomédicales* is on microfiche.

Guides to the literature

Guides to the literature soon date. Details of historical collections and museum specimens of plants and animals probably have a longer shelf life, but the upheavals in science funding in many countries have meant that many collections have been moved, or worse. The publishing world has also been in flux.

The subject guides of many libraries are often uncritical complications which provide a map of what is available: few produce guides to the most efficient use of the material. Most, if not all, libraries have a reference collection which contains guides to the literature. General guides to disciplines related to the life sciences are: *Information Sources in Chemistry*, edited by R.T. Bottle and J.F.B. Rowland, (4th edn., 1993), *Information Sources in the Medical Sciences*, edited by L.T. Morton and S. Godbolt (4th edn., 1992) and *Information Sources in Patents*, by C.P. Auger (1992), all published by Bowker-Saur.

The premier guide to science and technology is *Walford's Guide to Reference Material*, edited by M. Mullay and P. Schlicke (6th edn., Vol 1, Library Association, London, 1993) with 7000 entries covering books, encyclopaedias and secondary services. It is worth checking one's interests in such a compendium.

Information Resources in Toxicology, by P. Wexler (2nd edn., Elsevier, 1988), although old, is a comprehensive reference book containing information about pesticides, education, international bodies, legislation, computerized information sources etc. There are chapters for many countries, e.g. Argentina, Egypt, India and Japan, and information about food and other environmental matters, for example *Keyguide to Information Sources in Aquaculture*, by D.A. Turnbull (Mansell, 1989) is now very dated: for example it includes a Directory of Scientists in SE Asia Region from 1979.

Other resources

There are many resources for particular subjects. Experts may be located through research newsletters (p. 27). *A Directory for Entomologists*, by M. Colvin and D. Reavy (2nd edn, AES Publications, Colchester, 1993) is a guide for the UK with information about supplies, museums, local societies, reference libraries, experts etc. Specialized departments may have important collections of reprints, data and visual aids: the Hope Department of Entomology at Oxford, UK, has 52 000 reprints and 2000 slides. There is now a Centre for Insect Biology Research at the University of Arizona, Tucson, USA, funded by the National Science Foundation (NSF). The Natural Enemy databank, International Institute of Biological Control, Ascot, UK, is a resource for information about the natural enemies of arthropod pests. The Expert Center for Taxonomic Identification, ETI (Dr P.H. Schalk, Box 4766, 1009 AT Amsterdam, The Netherlands) has drawings, films, photos, maps etc. of marine organisms, and the Instituut voor Taxonomische Zoologie aft Entomologie, Plantage Middeniaan 64, Amsterdam, might be a resource for others. The former produces World Biodiversity Database, a series of CD-ROMs covering many groups of animals and plants, e.g. pelagic molluscs, marine planarians, sea mammals, birds of Europe etc. (see p. 208).

Finding people

Current Contents lists the addresses of authors and is a good starting place if one finds a recent paper. If your library has *Science Citation Index* you can use the author and institutional indexes. The geographical index is generated by a computer, which lacks the finer points of geography: some entries for Aberdeen are listed under England. For directories for biotechnology see p. 140 and for botany and agriculture see p. 254.

Current Research in Britain: Biological Sciences is published annually by The British Library and is available both in print and on CD-ROM. One can use it to identify those working on a particular topic. The *Directory of UK Biological Societies* is a guide to more than 70 societies affiliated to the Institute of Biology, London; the latest edition is 1996, but updates may be expected. The annual CD-ROM of *European R&D* Database lists 100 000 researchers in basic and applied science in 23 000 organizations in 37 European countries. It is also available in two printed versions, *Directory of European R & D* and *Who's Who in European R & D*, both annual.

World Research Database is a CD-ROM of 15 databases, including *Agriculture Research Centres*, *Medical Research Centres*, *Pacific Research Centres*, *European Research Centres*, *Industrial Research in the UK*, *Who's Who in Science in Europe*, *Agricultural and Veterinary Sciences International Who's Who*, *Medical Sciences International Who's Who*: clearly, life sciences is not too important! There are details of nearly 30 000 research centres, names and addresses of 70 000 senior scientists and 135 000 names and addresses. Although there is a *Pacific Research Centres* it is not clear how complete the cover is for South and Central America and the Indian subcontinent. There are regular updates by disc replacement. *The World of Learning* annual (47th edn., 1997, Europa) is up to date and good, but life scientists form only a small proportion of the entries.

Research and Development in Russia (RDIR) is a CD-ROM covering 4500 R & D institutes of the former USSR. There are more than 25 000 entries, including 15 000 abstracts of dissertations and reports.

Some societies produce annual lists of members with addresses, but others produce them at intervals. Some research newsletters print directories of their invisible colleges, and others would doubtless forward queries etc to the right person (see p. 27). Unfortunately, lists of members can be sold to advertisers etc., so that there is a financial loss if they are distributed free. The largest single life science membership society is the American Society of Microbiology, with 41 000 members and publishing an annual *Membership Directory.*

For medical doctors and other professionals who have to be registered, there are annual directories and searching is usually easy. Libraries, however, seldom have recent issues for foreign countries and one may have to seek the help of the relevant Embassy.

CHAPTER SEVEN

Reference works, dictionaries and encyclopaedias

H.V. WYATT

It is often difficult to know exactly what a new term or name means, and your library may not have a recent dictionary. Dictionaries date very quickly, especially in subjects such as immunology, and only current ones are listed. The following shows the range and includes a number of reference books.

International Dictionary of Medicine and Biology, (3 vols, Wiley, 1986). *Chambers Biology Dictionary*, edited by P.M.B. Walker (CUP, 1989) contains 10 000 definitions, including suffixes and prefixes, compiled from the database of the *Chambers Science and Technology Dictionary* which has 45 000 entries. *Dictionary of Gene Technology*, by G. Kahl (VCH, 1995) explains more than 4000 terms, with the help of 215 figures. It is especially good for unravelling acronyms. *The Encyclopedia of Molecular Biology*, edited by J. Kendrew (Blackwell, 1994) is illustrated and has 1160 pages. Reviews have ranged from too expensive to essential. *Biotechnology from A to Z*, edited by W. Bains (OUP, 1993) has 250 short entries from a half to three pages, in which over 1000 terms are explained. *Cell and Molecular Biology: Essential Data*, by D. Rickwood and D. Patel (Wiley, 1995) is inexpensive and has basic data on small and macromolecules, radioactivity, buffers etc., and should be a handy laboratory reference. *A–Z of Clinical Chemistry – a Guide for the Trainee*, by W. Hood (MTP, 1980) is old but still useful for very basic concepts.

Molecular Biology and Biotechnology, edited by R.A. Meyers (VCH, 1995) has over 1000 pages and 500 diagrams, with more than 250 articles and 1600 keywords. *Glossary of Biochemistry and Molecular Biology*, by D.M. Glick (Raven Press, 1990). *Dictionary of Biochemistry and Molecular Biology*, edited by A.D.

Smith (OUP, 1996) contains 15 000 entries, including 850 structures of organic and natural products and structural information on about 1500 proteins, with the filenames in a sequence database. *Enzyme Nomenclature 1992* (AP, 1992). *The Dictionary of Cell Biology*, edited by J.M. Lackie and J.A.T. Dow (2nd edn., AP, 1995) now has 5000 entries. There are tables of Cluster Differentiation (CD) antigens, restriction endonucleases, oncogenes, polypeptide hormones and growth factors.

Ainsworth & Bisby's Dictionary of the Fungi, edited by G.C.

How to do it, edited by D. Reece (3rd edn., BMJ, 1995) is a reprint in three volumes of a series of short articles from the *British Medical Journal*. Although there are many articles that deal with specifically medical situations, there are many very useful, well written and often humorous articles from which we could benefit:
Volume 1. Management/Employment/Counselling. e.g. How to chair a committee, give a reference, prepare a CV.
Volume 2. Examinations/Academia/Research and Communication. e.g. How to look after a visiting speaker, prepare for a foreign fellow.
Volume 3. Writing/Publishing and Publicity/Something Different.

Scientific Style and Format. The CBE Manual for Authors, Editors, and Publishers, Style Manual Committee of the Council of Biology Editors (6th edn., CUP 1994) is now very large, at 825 pages, and covers all the sciences. It is a mine of useful information, particularly for special scientific conventions. These cover all aspects of science, including 14 pages on viruses, as well as topics such as nomenclature for genes, viruses, fungi, plants etc., together with, for example, chromosome components, oncogenes, plasmids, initiation and elongation factors. There are many references, both cited and additional. There is a short list of culture collections, with addresses. Bibliographies on matters about journals and editing appear in *EASE Bulletin* of the European Association of Science Editors, together with articles and papers from conferences. Although the Association is mainly European, it has many members from around the world and welcomes any opportunity to further good practices.

Ainsworth *et al.* (8th edn., CAB International, 1995) has nearly 18 000 entries and includes lichens and yeasts. This is mainly a taxonomic work, not the usual kind of dictionary. *Dictionary of Microbiology and Molecular Biology*, by P. Singleton and D. Sainsbury (2nd edn., Wiley, 1988.) is now rather dated but still contains much of interest, and a new edition is due. There are 1000 pages and an appendix of microbial metabolic pathways. *Virology – Directory and Dictionary*, edited by R. Hull, F. Brown and C. Payne (Macmillan, 1988) is due for a new edition but has the following very useful appendices: Insect species, with infections by occluded baculovirus and granulosis, by cytoplasmic polyhedrosis virus, by densovirus, by entomopoxvirus, and by iridescent virus. Appendix F is a very long list of the properties of phage particles, alphabetically by phage. My copy lacks the list of references to Appendix F. See also *Virology Labfax* p. 187. *The Dictionary of Virology*, by B. Mahy (2nd edn., AP, 1995) is about clinical virology.

Dictionary of Immunology, edited by F.S. Rosen, L. Steiner and E. Unanue (Elsevier, 1989) has a useful list of abbreviations and a short list of CD antigens. *The Dictionary of Immunology*, edited by W.J. Herbert, P. Wilkinson and P. Scott (4th edn., AP, 1995)

Dictionary of Cytokines, by H. Ibelgaufts (VCH, 1994) is a 'unique review in dictionary form' with 3200 entries, 14 000 references and many coloured figures. It contains procedures and processes and seeks to explain the many synonyms and acronyms. This book will be useful not only to immunologists, but also to cell biologists and biochemists.

Glossary of Genetics, Classical and Molecular, edited by R. Rieger, A. Michaelis and M.M. Green (5th edn., Springer Verlag, 1991). *Encyclopedic Dictionary of Genetics*, edited by R.C. King and W.D. Stansfield (OUP, 1990) is a dictionary of genetics in English with each term given its German equivalent. There is a German/English index. *Essential Genetics*, by A. Hodson (Bloomsbury, 1994): the second half of this book is an encyclopaedia and dictionary (see p. 117).

Dictionary of Theoretical Concepts in Biology, edited by K.E. Roe and R.G. Frederick (London: Scarecrow Press; New Jersey: Metuchen, 1981). Despite its name this is no ordinary dictionary but instead is a source of references to the original paper where a concept was first described, and to key references since, e.g. Red Queen hypothesis. There is cross-referencing when several names are associated with a concept. *Concise Oxford Dictionary of Zoology*, edited by M. Allaby (OUP, 1991). *Concise Oxford Dictionary of Botany*, edited by M. Allaby (OUP, 1992) has 5000 entries. *A Dictionary of Botany*, edited by R.J. Little and C.E.

Jones (Van Nostrand, 1980) has 5500 terms and is illustrated. *Oxford Dictionary of Natural History.*, edited by M. Allaby (OUP, 1986). *World Dictionary of Livestock Breeds*, by I.L. Mason (3rd edn., CAB International, 1988).

A Dictionary of Plant Pathology, by P. Holliday (CUP, 1989) has 8000 entries with brief descriptions, taxonomic details and references to major texts and journal articles. There are biographical details of many phytopathologists and a bibliography of nearly 200 major phytopathological texts. There are details of nearly 700 proven viruses and others still in doubt. The author worked with the CAB International Mycological Institute at Kew, and this book is particularly good for its worldwide cover. *Dictionary of Environmental Science and Technology*, by A. Porteus (revised edn., Wiley, 1992) includes a list of organizations.

Dictionary of Abbreviations in Medical Sciences, by R. Heister (Springer Verlag, 1989) has 20 000 entries and a list of abbreviations of 500 journals. *Dictionary of Biomedical Acronyms and Abbreviations*, by J. Dupayrat (2nd edn., Wiley, 1990). *Dictionary of Drugs* (Chapman & Hall, 1990) has chemical data, structures and bibliographies 6000 drugs to July 1989, much from patents.

The Dictionary of Toxicology, edited by E. Hodgson, R. Mailman and J. Chambers (2nd edn., 1995). Chemical entries have CAS numbers and there are details of pertinent regulations, organizations, government bodies and associations in North America and Western Europe.

Encyclopedia of Antibiotics, by J.S. Glasby (3rd edn., Wiley, 1992).

AntiBase (not to be confused with Antisense) is a database of 17 000 natural antibiotics from microorganisms and higher fungi compiled, with regular updates, by Professor Laatsch of Gottingen University, available through Chemical Concepts, D-69469 Weinheim, Germany. *Dictionary of Alkaloids*, edited by I. Southon and J. Buckingham (Chapman & Hall, 1989) lists 10 000 alkaloids under 5000 entries.

Stedman publish many medical dictionaries and a series; *Stedman's ASP Parasite Names* 1993, with 3000 entries; *Stedman's ATCC Fungus Names* 1993, with 3000 valid names; *Stedman/Bergey Bacteria Words* 1992, with 11 000 names; *Stedman's ICTV Virus Words* 1992, with 5500 entries; *Stedman's Pathology and Lab Medicine Words* 1992, with 55 000 entries. All contain much additional information.

Illustrated Dictionary of Immunology, by J.M. Crase and R.E. Lewis (CRC, 1995) is wide-ranging with good illustrations.

CHAPTER EIGHT

Methods, products and quality control

H.V. WYATT

The past decade has seen the almost universal application of molecular and immunological techniques throughout the life sciences. At the same time, the interval between publication of a method and its adoption using commercial materials has become very short. At a virology conference held only 18 months after the description of the PCR (polymerase chain reaction) method, nearly a third of the papers presented were based on such applications.

There are books, manuals and journals as well as innumerable articles scattered through a very wide literature – some of the most cited papers are to what were new methods. One of the most useful aspects of *Science Citation Index* is the easy route to finding improvements and modifications to a published method. Some publishers have series, e.g. the *Benchtop Manuals* and *Biological Techniques* series by Academic Press and *The Practical Approach* series by IRL. The last-named claims to have sold half a million copies of 65 titles. From 1996 these have been available on a CD-ROM with annual update. Any ten titles are also available on a cheaper disc.

The best-known series is *Methods in Enzymology* (AP), with more than 276 volumes. *Methods in Enzymology on CD-ROM 1995* consists of 18 core volumes on recombinant DNA, with indexes to volumes 1–244. Depending on the sales of this compilation we may see further developments, both in this series and in others such as *Methods in Microbiology.*

A new development is the simultaneous publication of *Current Protocols in Immunology* in a looseleaf two-volume format and a CD-ROM version, 1991 Wiley. Supplements are available twice a year for the paper version and four times for CD-ROM. There is also a newsletter, with suggestions from users. There is a chapter on animal care and handling. A similar publication from Wiley is *Current Protocols in Molecular Biology*, for which there is a core looseleaf manual, *Short Protocols in Molecular Biology*, (2nd edn, Wiley, 1990) edited by F.M. Ansubel *et al.* The suppliers listed in these books are for the USA, not Europe (see instead *Linscotts Directory*, p. 44. A feature of the CD-ROM is that protocols may be modified by the user and imported into wordprocessors and printed for use in the laboratory.

Methods of Immunological Analysis, edited by R.F. Masseyeff, W.H. Albert and N.A. Staines, 11 volumes 1992–1997, plus an index volume, is a massive reference work from VCH. The first three volumes cover basic knowledge of immunology.

There are Videos for Science demonstrating molecular biology techniques, including cell culture, PCR and recombinant DNA, from Taped Technologies, University of Utah, USA.

There are many books on methods, some of which are given in the various chapters of this book. One of universal application is *Immunoassay Automation: a Practical Guide*, edited by D.W. Chan (AP, 1992).

Journals Devoted to Methods only:

Analytical Biochemistry includes molecular genetics, immunoassays, cell and organ culture, with reviews and 'kitchen tricks';
Genetic Analysis has reviews of new technologies;
Journal of Tissue Culture Methods 1975– TCA, Gaithersburg MD;
Journal of Immunological Methods;
Journal of Microbiological Methods;
Journal of Virological Methods;
Methods 1990– is described as a companion to *Methods in Enzymology*, with each issue devoted to a single topic. There are also
Methods in Cell Science 1978–;
ImmunoMethods 1992–;
Neuroprotocols 1992– and
Genomethods 1995– which are similar journals.
Technique 1989– is about methods in cell and molecular biology, with good illustrations and clear separate protocols.
PCR Methods and Applications.

Manufacturers

Many manufacturers, such as Bio-Rad, provide or sell booklets about specific techniques, which are advertised in the free laboratory newsletters. Sigma Diagnostics market 100 books, of which five are on immunology and 14 on microbiology, with many others on haematology and histology etc.: this is a unique collection of the best books for the laboratory.

Methods and apparatus often fail to work as well as they should in a new laboratory. Sometimes the reagents come from a different source and may contain an unexpected impurity, or lack something which is vital; sometimes the instructions are not as clear as they should be. It is always good to visit a laboratory where the method is already in use and to watch it in practice. Remember to take home samples of the reagents. Friends can be useful, or the manufacturer may be able to enlist the help of another customer.

Products

There is an annual UK Product Review Supplement to *Nature*, as well as special product reviews, usually subject based, in most issues: the advertisements are also very useful. *Science* has an annual instrumentation issue with articles about techniques and apparatus, and special issues, for example as trends in analytical instrumentation. *Biotechnology* has supplements called *Europroduct Focus*, which are largely advertisements but also have interesting reviews. The autumn 1995 *Focus* had a review of 41 biological CD-ROMs and 27 fermenters and bioreactors. Each review was in the form of a comparative table.

In addition to the advertisements, most of the trade journals have become more like review journals, with articles on techniques, safety, laboratory management etc. Although most are aimed at clinical laboratories, with their expensive and continuous running costs, there is much of interest for the researcher and academic.

Many suppliers have informative booklets about techniques, organisms etc., and many are free. Sigma Diagnostics sells a range of 100 books on laboratory tests, stains etc. Some catalogues are particularly useful, with large sections of reference material about their products. New England BioLabs, which supplies biotechnology enzymes, kits etc., has a 90-page reference appendix with restriction maps, cloning strategies, properties of restriction enzymes, genomic analysis etc. – information which would be difficult to obtain from any other source. The same firm publishes a newsletter, *The NEB Transcript*, with short informative articles, including those by the Nobel Laureate Dr R. Roberts. A commercial newsletter, *Transduction Laboratories Insights* 1995–, from the specialist supplier Affiniti Research Products, covers signal transduction: the notes, however, show that the important papers come from a limited few journals.

Linscotts Directory of Immunological and Biological Reagents (9th edn., 1996–1997) is in both printed and disk versions. About 60 000 products are listed as monoclonals, polyclonals, miscellaneous, organisms and kits, with many subdivisions within each. There are two product indexes, one for antibodies and hybridomas and another for everything else. Against each product are listed the suppliers: there is an index of the 450 suppliers, together with their 1500 distributors. Supplements are issued every six months between editions (every two years). Each disk supplement, both DOS and Windows versions, is a complete new edition.

Trade journals

American Clinical Products Review
Clinical Laboratory International, with company profiles
Life Science
European Clinical Laboratory, with both American and European Buyers' Guide Editions
Biotech Products International does not cover the USA
Microbiology Europe (formerly *European Microbiology*) has book reviews, articles and news from Eastern Europe
Laboratory Products International is aimed at Europe and the Asia-Pacific areas, and has an Environmental News Supplement
Laboratory Products Technology, including *Life Science News*
Lab Medica International
Laboratory News, with supplements on biotechnology, microbiology, microscopy and diagnostics, is one of the largest of the business publications, with many articles as well as advertisements
Biotech Literature Review and *Laboratory Literature Review* are simply advertisements of catalogues etc.
Scientific Computing World 1994– contains the usual advertisements for statistical software, but also many articles, e.g. on automating DNA analysis in September 1995, and Bioinformatics: computers at the heart of molecular biology in October 1995.

Diagnostics in Africa is distributed with *Africa Health* and contains short articles, news and letters etc.

Standards

Standards are important in all scientific work and are required for clinical and pharmaceutical preparations and tests. When a laboratory first enters a quality control scheme, about half of the routine tests are often unsatisfactory and it is hard work to make improvements. Sometimes common errors are revealed and methodology can be changed. Schemes should be independent of reagent or manufacturing interests. Research and academic laboratories seldom have quality tests, and in many cases there will be no suitable outside schemes.

Commercial kits themselves may not be as accurate as claimed: in 1993 the French government tested 31 HIV kits and found nine to be unsatisfactory. Some years ago, comparative tests were arranged for a routine assay of polioviruses at two temperatures. The tests showed that the water baths used in different laboratories were not properly controlled and previous work was probably invalid. With computer networking it should be easier for research groups to arrange cooperative quality testing and standards.

Sigma Diagnostics supply ACCUTROL control reagents with both normal and abnormal levels for their products, and a conversion chart for all other major manufacturers. Sigma offers a free quality verification service for Enzyme immuno-assay (EIA). Other manufacturers with control standards include Bio-Rad, Difco, Miles and Roche. For microbiological media, control cultures from the American Type Culture Collection (ATCC) are often listed. Microbial contamination is a great problem in industry, and an estimated one and a half billion tests are made each year: bioluminescence kits are a new development (Celsis of Cambridge, UK).

Tissue cultures are especially liable to contamination by low-grade infections with either mycoplasma or bacteria. The routine inclusion of antibiotics and antifungal agents may conceal low-grade contamination, and long-term cultivation of samples at two different temperatures without these agents is recommended. There has been much cross-contamination by HeLa cells, which insidiously replace the original cell line: special care must be taken if other cell lines are used in the same laboratory. There are standard tests for specific enzymes and antigens which should be routinely made on all cultures before and after a piece of work is done. Although journals do not insist on full details of the origin of cell lines, or on the results of routine tests for contamination, articles should be carely scrutinized for these details and all authors should include this information in their publications. The *ATCC Quality Control Methods for Cell Lines* (2nd edn., 1992) should be in every tissue culture laboratory (see also p. 150). Difco and Oxoid sell excellent books on their media, (see p. 181).

In the USA the National Committee for Clinical Laboratory Standards (NCCLS) publishes standards, methods and guidelines through a series of expert committees. Subjects include microbiology, molecular methods, immunology, toxicology, chemistry, safety etc. The Unit of Health Laboratory Technology publishes manuals, guidelines, reports etc. on a wide range of clinical techniques, including microbiology and immunology; lists are published in *Journal of the International Federation of Clinical Chemistry.*

The National Institute for Biological Standards and Control (NIBSC) in the UK produces primary international reference materials for biological substances: other centres may be found from WHO. The international Steroid Reference Collection is held by Sigma, which also hosts the Sigma–Aldrich Library of 40,000 research samples, with a structure index to it.

The US Food and Drug Administration (FDA) is responsible for licensing products in the USA, which is probably the biggest and most important market in the world. It is therefore necessary to comply with FDA regulations and good practices. Federal and State laws demand certification, accreditation etc.: *Laboratory Regulation News* every two weeks gives news of US and foreign, including European Union, requirements (see also p. 00). A reference book is *Good Laboratory and Clinical Practices. Techniques for the Quality Assurance Professional*, edited by P. Carson and N. Dent (Butterworth–Heinemann, 1994).

The American Organization for Analytical Chemistry (AOAC) works with similar international bodies to validate standard methods, with a network of over 1000 members. *Official Methods of Analysis of the AOAC*, by K. Helrich (16th edn., 2 vols, looseleaf or as a CD-ROM, 1995) contains more than 2300 methods for chemical and microbiological analysis and covers drugs, pesticide formulations, foods, flavours, additives etc. There will be five annual supplements. *Analytical Abstracts (ANAB)* includes clinical, biochemical, pharmaceutical, food, agricultural and environmental analyses, and is based on 250 journals as well as books, standards, technical reports and conference proceedings. It may be accessed online.

External quality assessment

In Europe, 23 organizers of external quality assessment schemes (EQAS) set up *EQAnews* in 1990 and form a focus for schemes worldwide. The newsletter goes to more than 250 people worldwide. Supplement 212 of the *Scandinavian Journal of Clinical and Laboratory Investigation* 1993 reports on a meeting on EQAs held in 1991. A list of addresses of nearly 200 organizers worldwide was printed in *EQAnews* 1994; 5 (2). The European Society of Clinical Virology, which also has a newsletter, has a quality assessment committee and organizes quality control exercises. It is seeking links with similar groups, such as the Panamerican Group for Rapid Viral Diagnosis. The first European conference on quality control in clinical laboratories was held in November 1995, and a number of schemes were discussed. The Commission of the European Communities is considering a European quality management system for in vitro diagnostics.

The UK National External Assessment Scheme (UKNEQAS, Box 401, Sheffield S5 7YZ) is available to public and private sector health as well as non-clinical laboratories, and has participants from all over the world. Schemes of interest to life scientists include, among others, viruses, immunology, cytogenetics, serology, bacteria, blood groups, HLA etc. The microbiology scheme has 500 participating laboratories in the UK and over 200 from 31 other countries. There are subschemes for antibiotic assay and parasitology. There are also schemes for water microbiology and food microbiology.

Bio-Rad operates a free quality control program for up to three years, with values and comparisons with other participating laboratories. Murex Diagnostics has EQAs in clinical chemistry, e.g. urea, creatine kinase, drug monitoring (e.g. antibiotics) and immunoassay (e.g. insulin, renin, immunoglobulins). The schemes operate worldwide.

The College of American Pathologists submits 16 faecal specimens to laboratories each year: correct identifications for helminths range from 80 to 94%, and for protozoa much less. A survey by Isolab Inc. of Akron, Ohio, in 1991, of 414 laboratories for a GHb sample showed 26 different methods from 22 different companies (*Clinical Laboratory International* June 1991).

CHAPTER NINE

Statistics and software

H.V. WYATT

'Make haste to use a new remedy before it is too late.' (Traditional)

Occasional analyses of papers in high-ranking large-circulation medical journals have shown that in spite of very careful refereeing, and selection of only about 10% of submitted papers, many selected papers are deficient in statistical validity.

Authors should never use 'not significant', but should always give the *P* value. Authors seldom give the null hypothesis which they are testing, nor is it always clear which of several possible hypotheses is being tested. Medical journals now require confidence limits to be calculated and given. Excellent books which are easy to read and understand are: *Statistics at Square One*, by T.D.V Swinscow (9th edn., BMJ, 1996); *Statistics With Confidence – Confidence Intervals and Statistical Guidelines*, edited by M.J. Gardner and D.G. Altman (BMJ, 1989), for which there is a computer program; and *Confidence Interval Analysis*, by M.J. Gardner, S.B. Gardner and P.D. Winter, (BMJ, 1989).

Although the advice may be difficult to carry out, it is always wise to consult a statistician before embarking on a study which may depend on a statistical interpretation. *Biometry* by R.R. Sokal and F.J. Rohlf (3rd edn., Freeman, 1995) has a key to statistical methods as the end paper. See also p. 231.

Both the *BMJ* and *The Lancet* have occasional series on statistics for the non-statistician. Nine of these pieces were reprinted as *Facts, Figures, and Fallacies. Epidemiological Pitfalls in Medical Research* (The Lancet, 1994) covers very basic topics not usually found in textbooks; however, the little book of 28 pages is very expensive and the articles can be read in the original (see, for example, Jolley, 1993).

A useful reference book is *The Cambridge Dictionary of Statistics in the Medical Sciences*, by B.S. Everitt, (CUP, 1995).

There are several computer programs for statistics. Some are designed for the life sciences, some not. Although some journals have accepted three-dimensional effect histograms prepared on programs designed for advertising use, the data are difficult to relate to the axes and would be clearer in two dimensions. Such an example is used in an advertisement for one product which 'makes your data look terrific'. Users may put the data into unsuitable programs: I have seen data only suitable for a histogram presented as a meaningless computer-generated 'graph'. The advertising booklet for a well-known 'statistical analysis program' shows a graph with placebo and two hormones as the axes.

Reviews of computer software appear in *Quarterly Review of Biology*, for example detailed reviews and comparative evaluation of six statistics programs for the Macintosh (*Quarterly Review of Biology* 1993; 68: 637–648). These reviews are particularly useful as there is a directory of publishers with telephone numbers; there is a similar directory for the books section. Other reviews appear in some of the monthlies, e.g. *TIBS* and *BioEssays*. Synopses, but not reviews, of new software programs are noted in *Biological Abstracts/RRM*. *Computer Applications in the Life Sciences*, also known as *CABIOS*, is a quarterly journal with original papers and reviews of software. *Biotechnology Software Journal* has software reviews of all kinds, and has an annual index. More educational software is reviewed in *Biologist*, *Journal of Biological Education* and *Biochemical Education*. *Binary – Computing in Microbiology* ceased publication at the end of 1995, but may emerge in a different form on the World Wide Web.

The Computers in Teaching Initiative (CTI) Centre for Biology at Liverpool University, UK produces *Life Sciences Educational Computing*, an occasional newsletter with news, addresses, reviews and short articles; it has lists of software vendors, including some from the USA. The major part of the newsletter is on the World Wide Web, http://www.liv.ac.uk/ctibiol.h+m/. A similar centre in the USA is the BioQUEST Curriculum Consortium at Beloit College, Wisconsin, which circulates *BioQUEST Notes*. Many of the molecular databases supply specialized software for use with their sequence data (see p. 142). Educational software for sequence, statistical and other applications may be useful.

Cherwell Scientific, Oxford, UK, sells *Cyrillic* for pedigree drawing. IRL Press specializes in software for teaching and research in the life sciences. Biosoft sells numerous programs and publishes the free *Biosoft News. Software for Science* offers over 2500 programs and CD-ROMs in the whole of the sciences, from SciTech International, IL 60614–3011, or Software for Science SG6 2HP in the UK. Megalon sells software for scientists: statistical analysis and presentation, chemical formulae, data and structures etc.

Programs for managing references and bibliographies are advertised in the weeklies and monthlies, e.g. *EndNote Plus*, *Reference Manager*, *Pro-Cite* and *Biblio-Link II* (see p. 89 for a comparison of several systems).

Programs for statistics, graphics and bibliographical applications are advertised in the weeklies such as *Nature* and *Science. Technical Computing* is a free magazine with advertisements and articles.

Reference

Jolley D. The glitter of the *t* table. *Lancet* 1993; 342: 27–29.

CHAPTER TEN

Journals: searching and impact factors

H.V. WYATT

> If you make a subject you never have to learn it. (S. Brenner, BBC talk 1987)

Chemists are likely to search *Chemical Abstracts*, whereas life scientists have a bewildering choice of secondary sources. A biochemist will have the choice of *Chemical Abstracts*, *Index Medicus*, *Biological Abstracts*, the *CABS* series. *Excerpta Medica*, as well as specialist serials such as the CABI serials. Each is prepared from a different selection of journals, with different delays. There have been few studies that compare searches using different databases and sources (see p. 55). Each of us will get used to the most readily available, and many will have no choice.

It is always valuable to look around the library of another institution to see what secondary serials, newsletters etc. are available. In particular, one should look for the bibliography section for items from unlikely sources: in some libraries bibliographies may be shelved with books. It is always useful to look at the books and the recently received journals for one's subject. Today visiting is more difficult, as access may be limited by library card or other electronic means: ask and you may be admitted. However, one survey showed that even visiting a library is nowadays not obligatory and many do not use secondary serials (Table 10.1).

Oceanographic Literature Review. Deepsea Research, Part B is an abstract serial which might not be the first choice for a microbiologist, yet it has a microbiology section. Some of the items are from mainstream journals, but there are items from relatively obscure journals, in Russian and Spanish, for example.

Many of the online databases may be searched through a

Table 10.1. Survey of use of library services by biologists

		Academic staff	Research students
n		27	9
Regular use of library		81%	100%
Sources of information:	colleagues	63%	78%
	other	93%	100%
	online	19%	11%
Use of abstract journals:	regular	48%	33%
	infrequent	15%	0%
	not at all	37%	67%

(Courtesy of Professor A.J. Meadows)

number of hosts, such as DIALOG, DIMDI, ORBIT etc. The choice depends on the particular needs of the researcher, who should consult an expert in the library. The various hosts may package the same database in different ways and charge at different rates, according to how the packages are used.

The *Directory of Online Databases* has two issues and two supplements a year, and lists thousands of databases available through hundreds of services. Most of the services, including databases, are now available as printed versions, online, as disks and as CD-ROMs. The online versions may be updated each night or weekly. Disks are updated more rapidly than CD-ROMs, for example *Current Contents, Medical Science Weekly* and *Reference Update*

Finding all the references

A project to locate all reference to costs and the economics of different methods of storage and retrieval of information estimated that there might be as many as 7500 references. Forty sources were searched and 6100 references were found in 10 450 locations. Only 5% were found by online searches, the most expensive method. Twice as many were found in a single serial, *Referativnyi Zhurnal Informatika*. Twenty-one percent were found from bibliographies – the best source – followed by card indexes 17% (British Library Research Paper no 37). Although these results may not apply to specific subjects in the life sciences, it is probable that the results are generally applicable. However, few scientists will do such an exhaustive search unless they are applying for a patent. Most scientists will assume that the significant research will have been reported in a relatively few journals (see p. 55).

Table 10.2. A search of six databases for two subjects in parasitology 1984–16 August 1988. Adapted from Ayres and Taylor (1988)

Subject	Number of items retrieved from					
	Public Health & Tropical Medicine	SciSearch	EMBASE	BIOSIS *Previews*	Medline	CABI
Strongyloides/						
Strongyloidiasis,	98	173	230	360	385	473
of which unique papers	1	1	8	20	29	67
Cryptosporidium/						
Cryptosporidiosis,	227	490	347	466	509	550
of which unique papers	4	7	8	11	10	50
Papers which appeared in	2	3	4	5	6 databases	
Strongyloides/						
Strongyloidiasis	34	13	17	9	6	
Cryptosporidium/						
Cryptosporidiosis	32	31	16	18	18	

on disk are all updated weekly, whereas CD-ROMs are often updated monthly.

A search of four databases yielded over 600 references: CABI 421 references, BIOSIS 107, AGRIS 105 and AGRICOLA 33. Other sources, e.g. *Tropical Agriculture Abstracts* and library subject indexes produced more references. The overlap of journal titles covered by *Index Medicus* and *Excerpta Medica* is only 36–40%, and the latter contains far more European and Japanese literature.

A similar search was published in *Parasitology Today* (Table 10.2). *Public Health and Tropical Medicine* only reviewed what were considered to be important papers: others, even if found in the journals covered, would be ignored. It is surprising that several papers considered important were found only by this service. It is also surprising that the most papers were found in the CABI database, surely not the first choice which comes to mind. It has always been clear that, although the most important literature may be found in several databases, there are important differences between the various secondary services, especially if one is interested in aspects that diverge from what are seen as the major research fronts.

What are judged the most important papers will probably appear in a few key journals and these are likely to be indexed or abstracted in several of the major secondary serials. K.S. Warren showed that 10 286 papers on schistosomiasis from about 1850 onwards were published in 1738 journals, but that one-third appeared in only 19 journals and a half in 50 journals. There are

Table 10.3. The journal literature of AIDS and HIV compared with some microbiological subjects

Subject	Period	Months	Number of Journals	Number of Papers	Average no. of papers/journal	% papers in top 10 journals
AIDS and HIV	1985	21	348	2561	7.3	46
	1987	15	446	3492	7.8	41
	1994	12	402	3220	8.0	40
Medical bacteriology	1980	6	521	1744	3.3	14
Cell structure & function	1980	6	235	1802	7.7	43
Immunology	1980	6	253	663	2.6	31
Vaccines	1980	6	34	48	1.4	52

AIDS and HIV were taken from *Current AIDS Literature*, previously *AIDS and Retrovirus Update* (see also Table 10.4).
The other subjects are taken from Carson (1984), who analysed entries in *Microbiology Abstracts* for January to June 1980 inclusive.

few studies which show whether the papers produced by further searches are truly marginal or are actually of interest.

To search only for existing terms or concepts or authors and journals may be very limiting. The basis of the work on transplantation immunity was suggested by work on twin cattle. The work which led to discovery of the role of DNA in genetics was originally published in *Journal of Hygiene*, and followed up with Avery's paper in *Journal of Experimental Medicine*, which had no clues to its importance for genetics in either its title or its abstract, and not much in the paper itself (Wyatt, 1972a). It is also very important to go back to the original. The famous Blender experiment of Hershey and Chase has been mutilated and trivialized in textbooks and the numerical data have been remembered with advantages by academics (Wyatt, 1974). The original diagram illustrating bacterial growth was simplified many, many years ago, and the false timescale has been copied ever since.

Computer searches can produce unlooked-for results: for example, a search for 5HT receptors retrieved papers where the author's postcode contained 5HT.

As services such as *Index Medicus* are always selective, with few journals from developing countries, it is difficult to locate local problems and data in those countries. There have been several attempts to remedy this, with, for example, *Index Medicus SE Asia*, a similar one for Africa, *LILACS* for Latin American and Caribbean health sciences literature from PAHO, which includes books, reports and theses and *ExtraMED* (see p. 108). Other local serials are: *Quarterly Index to Periodical Literature; Eastern and Southern Africa* 1991–, (Library of Congress Oversea Acquisition Office, Nairobi), with sections on biomedicine and science (e.g. *Zimbabwe Science News*); *Current Health Information Zimbabwe* 1987 – with abstracts, but mainly from international journals; *Pest Management Documentation & Information System & Service (PMDISS) Bibliography* with abstracts 1991 – from the International Centre of Insect Physiology and Ecology ICIPE, Nairobi.

With the increasing availability of computers in developing countries, relatively small sums of aid money could be used to produce more serials and bibliographies like the above (see also Bioline, p. 18). Unfortunately, there is also an urgent need to preserve past literature in national libraries. Lack of money, wars and unrest, and the problems of heat, dust and destructive insects and moulds have left many countries with impoverished libraries and information resources. For a bibliography of poliomyelitis in India, I was unable to find many Indian papers in India and had to complete the work in the UK. Too often, data from temperate countries are assumed to be relevant to conditions in the tropics and developing countries, and data from them are lacking or cannot be traced.

HIV and AIDS: How the literature grows

> Science to me is the finest sport in the world (Sir Macfarlane Burnet, 1981)

There has been intense interest in HIV and AIDS and, despite the very large number of papers, the literature was very compact in the late 1980s (Table 10.3). The papers in this survey had already been selected for inclusion in the secondary serial, whereas *Microbiology Abstracts* is not selective except in the journals used. There are very considerable differences in the structures of various subjects within the umbrella of microbiology.

Another analysis shows the ranking of the top journals for AIDS and HIV, again from the selective secondary serial (Table 10.4). AIDS was not typical of an ordinary subject because the agent and the mode of infection were unknown at first. Because of the intense interest and speculation, there was concentration of papers in the weeklies, but specialist journals were slow to start. By 1996 there were specialist AIDS journals, seven of which were in the top 17 journals (Table 10.5). The weeklies were still near the top, but the literature now reflected other interests and

Table 10.4. Ranking of journals carrying papers and comments about AIDS and HIV. Analysis of items in *Current AIDS Literature*, previously *AIDS and Retrovirus Update*

<table>
<tr><th colspan="3">21 months October 1985–June 1987</th><th colspan="2"></th><th colspan="2">15 months July 1987–Sept 1988</th></tr>
<tr><th>Rank</th><th>%</th><th>Cumulative %</th><th colspan="2">Title</th><th>%</th><th>Cumul. %</th></tr>
<tr><td>1</td><td>10.7</td><td>10.7</td><td colspan="2">Lancet</td><td>9.6</td><td>9.6</td></tr>
<tr><td>2</td><td>6.1</td><td>16.8</td><td colspan="2">JAMA</td><td>4.9</td><td>14.5</td></tr>
<tr><td>3</td><td>5.8</td><td>22.6</td><td colspan="2">BMJ</td><td>4.4</td><td>18.9</td></tr>
<tr><td>4</td><td>4.5</td><td>27.1</td><td colspan="2">Ann Intern Med</td><td>4.1</td><td>23.0</td></tr>
<tr><td>5</td><td>4.4</td><td>31.5</td><td colspan="2">Nature</td><td>3.6</td><td>26.6</td></tr>
<tr><td>6</td><td>3.7</td><td>35.2</td><td>N Engl J Med</td><td>AIDS Forschung</td><td>3.6</td><td>30.2</td></tr>
<tr><td>7</td><td>3.1</td><td>38.3</td><td>Science</td><td>AIDS</td><td>3.2</td><td></td></tr>
<tr><td>8</td><td>2.7</td><td>41.0</td><td>AIDS Forschung</td><td>N Engl J Med</td><td></td><td>36.6</td></tr>
<tr><td>9</td><td>2.4</td><td>43.4</td><td>AIDS Research</td><td>Science</td><td>2.7</td><td>39.3</td></tr>
<tr><td>10</td><td>2.3</td><td>45.7</td><td>J Infect Dis</td><td>AIDS Res</td><td>2.1</td><td>41.4</td></tr>
<tr><td>11</td><td>1.4</td><td>47.1</td><td>Am J Med</td><td>J Virol</td><td>1.8</td><td>43.2</td></tr>
<tr><td>12</td><td>1.3</td><td>48.4</td><td>Proc Natl Acad Sci USA</td><td>J Infect Dis</td><td>1.5</td><td></td></tr>
<tr><td>13</td><td>1.2</td><td></td><td>Deutsche Med Wchsthr</td><td>Am J Med</td><td></td><td>46.2</td></tr>
<tr><td>14</td><td></td><td>50.8</td><td>Morb Mort Week Rep</td><td>Am J Pub Health</td><td>1.3</td><td></td></tr>
<tr><td>15</td><td></td><td></td><td></td><td>Proc Natl Acad Sci USA</td><td></td><td></td></tr>
<tr><td>16</td><td></td><td></td><td></td><td>Morb Mort Week Rep</td><td></td><td>50.1</td></tr>
<tr><td>Total:</td><td colspan="2">journal titles</td><td>348</td><td>446</td><td></td><td></td></tr>
<tr><td></td><td colspan="2">journal items</td><td>2561</td><td>3492</td><td></td><td></td></tr>
<tr><td colspan="3">Journals with 2 citations</td><td>55 (16%)</td><td>65 (14.5%)</td><td colspan="2">1.8% of citations</td></tr>
<tr><td></td><td colspan="2">1 citation</td><td>154 (44%)</td><td>213 (48%)</td><td colspan="2">6.1% of citations</td></tr>
</table>

priorities, such as care, education and paediatric cases. France, with the greatest number of cases in Europe was slow in tackling the disease, but a French-language journal was in twelfth place in 1994. As the epidemic has spread, other language journals have become important. Nevertheless, the literature is still remarkably compact (see Table 10.3).

The AIDS epidemic generated a new literature from 1981, with very short doubling times which then increased (Table 10.6). Such an enormous rise in the literature must be unprecedented.

A more normal picture of a new field would be the original discovery of polyoma virus, where the first papers appeared in a

Table 10.5. Ranking of journals cited in *Current AIDS Literature* Jan-Dec 1996

Rank	Journal	%	Cumul. %
1	*AIDS*	7.66	7.66
2	*J AIDS*	7.11	14.77
3	*AIDS Res Human Retroviruses*	6.64	21.41
4	*J Infect Dis*	3.99	
	J Virol	3.99	29.39
6	*Clin Infect Dis*	3.17	32.56
7	*Lancet*	2.45	35.01
8	*Virology*	2.36	37.37
9	*Intern J STD & AIDS*	1.81	39.18
10	*Med J Australia*	1.57	40.75
11	*AIDS Care*	1.51	42.26
12	*AIDS Educ Prevention*	1.34	
	Antimicrob Agents Chemother	1.34	44.94
14	*J Biol Chem*	1.22	
	Pediatr AIDS HIV Infect	1.22	
	PNAS	1.22	48.60
17	*N Engl J Med*	1.16	49.76

447 journals were cited of which 279 (62%) were cited more than once.
55 journals had titles which suggested that the majority of papers in them were not in English. 3434 papers were cited of which the probable language was

English	3272	95.28%	Dutch	11	0.32%
French	61	1.78%	Italian	10	0.29%
Spanish	37	1.08%	Eastern European		
German	35	1.02%	and Russian	7	0.20%

Original printout kindly supplied by D. W. FitzSimons.

variety of journals according to the tissue culture, immunology, virology and oncology findings.

How papers appeared in the secondary literature

Two of the most important aspects of finding information is where it is and when it becomes available. An important paper about a possible treatment for HIV infection appeared in *Science* in 1988. One of the drugs used was ddC (dideoxycytidine), which was used with zidovudine (AZT) in the successful Delta trial, the results of which were announced in September 1995. The *Science* paper (see Fig. 10.1) was noticed in the major secondary serials: of many papers examined, this was the only one to appear in all the serials scanned (Table 10.7). The paper was given a press release which was reported in the *International Herald Tribune* and formed the basis for item 400 in *AIDS Newsletter*. Before the copy of *Science* had arrived in some British libraries, it was the subject of a short article in *New Scientist*, followed a month later by a note in *AIDS*

Table 10.6. The growth of the AIDS and HIV literature as recorded by secondary services

	Source	Period	Doubling time, months
AIDS	*Index Medicus* and BIOSIS	1981–1984	4
		1985–1988	22
HIV	*Index Medicus*	1984–1987	3.5
		1987–1988	13

Note that the BIOSIS database included all HTL viruses, not just HIV.

Newsletter and in *ATIN*. An editorial comment followed in *Current AIDS Literature*. All these gave not the abstract or a summary, but details of its importance, possible use and action: all were useful.

The authors' abstract did not mention the use of ddC, zidovudine and ddA in their experiments, or the effect on reverse transcriptase and mammalian DNA polymerase, which were noted in *Current AIDS Literature* and *AIDS Newsletter* (Fig. 10.1). A search by specific names in Medline would have retrieved the paper, but the abstract would have been unhelpful (Fig. 10.1). The abstract was given in *Biological Abstracts* in small print, in *AIDS Information* in readable form, but slightly abridged, and in *AIDS Research Today* with added descriptors. *AIDS-Forschung*, which gives details of important papers as well as printing original papers, gave its own views, not the abstract.

The paper took about two months to appear on-line, more than three months on CD-ROM and five months to appear in print in *Index Medicus* (all times are UK availablity) (Table 10.7). It took even longer for *AIDS Research Today*, which is prepared from the BIOSIS database, to reach the UK.

The title of the paper contained the words suppression, viruses, inhibition and virion binding which like HIV are too general for searching with the KWIC index of *Biological Abstracts*. Dextran sulfate would be a keyword for very few, if any, researchers. The remaining term CD4 would be the obvious keyword for many researchers interested in HIV interactions with cells. However, the entry in the KWIC index (Fig. 10.1)

'virion binding to CD-4 positive cells human immunode' includes neither suppression nor inhibition. The print is difficult to read and the information is very limited and not very informative so that retrieval of the paper from *Biological Abstracts* is difficult. The printout from Medline was also disappointing: the

Authors' abstract:
Mitsuya H, et al. Dextran sulfate suppression of viruses in the HIV family: inhibition of virion binding to CD4-positive cells. *Science* 1988 (April 29) 240; 646–649.

The first step in the infection of human T lymphocytes by human immunodeficiency virus type 1 (HIV-1) is attachment to the target cell receptor, the CD4 antigen. This step may be vulnerable to attack by antibodies, chemicals, or small peptides. Dextran sulfate (molecular weight approximately 8000), which has been given to patients as an anticoagulant or antilipemic agent for more than two decades, was found to block the binding of virions to various target T lymphocytes, inhibit syncytia formation, and exert a potent inhibitory effect against HIV-1 in vitro at concentrations that may be clinically attainable in human beings. This drug also suppressed the replication of HIV-2 in vitro. These observations could have theoretical and clinical implications in the strategy to develop drugs against HIV types 1 and 2.

This abstract was given in *Biological Abstracts* and in *AIDS Research Today* (derived from the BIOSIS database): in the latter, some key words were also given. *AIDS Information* and *ATIN* gave a shortened version of the abstract. *ATIN*, however, gave the editor's comments:

ED's COMMENTS: Dextran sulfate is thought to inhibit binding of virus to cells. It also inhibits syncytia formation and may have an antiviral effect at concentrations achievable *in vivo* by oral administration. The data of Mitsya and coworkers confirms inhibition of HIV-1 binding to cell, but it is unclear whether this is due to disruption of the interaction of (a) the HIV-1 surface protein (SU) with T4 or (b) the HIV-12 transmembrane envelope protein (TM) with the cell membrane. The toxicity and efficacy of dextran sulfate will be studied in humans soon. However, pharmacological evaluation of this agent may be difficult.

AIDS-Forschung gave a commentary in German.

AIDS Newsletter gave two comments, one about the dextran sulfate, the other, from *New Scientist* about the other agents not mentioned in the authors' summary:

■ 399 ■ Dextran
Leading researchers at the National Cancer Institute in the USA have reported that dextran sulphate, a large sugar molecule, protected T lymphocytes from damage by HIV and also prevented infected cells from fusing with uninfected cells as they usually do without dextran. (This fusion is believed to be a key step in destroying the immune system.) It also blocked HIV from attaching to the surface of uninfected cells. The compound was already considered a high priority for study and these findings will fuel interest in it. Already it is the most popular "underground" treatment for AIDS.
International Herald Tribune, 30.4.88, p3

■ 400 ■ Antiviral agents
Samual Broder of the National Cancer Institute of Bethesda, Maryland (USA), is confident that HIV infection will be curable. Besides zidovudine and dideoxycytidine (ddC), he identifies as other potentially successful chemotherapeutic agents dideoxyadenosine (ddA, which has so far been given to 5 AIDS patients), phosphonoformate (Foscarnet), interferons, soluble CD4, acyclic nucleoside analogues, ampligen, and dextran sulphate. With his colleagues he has shown (see *Science*) that dextran inhibits both HIV-1 and HIV-2 but it also affect a key human enzyme for DNA synthesis and repair. They note synergy between dextran sulphate and some nucleoside analogues including zidovudine, ddC and ddA.
New Scientist, 26.5.88/H Mitsuya et al., Science, 1988, 240(Apr 29), 646-9

FIG. 10.1: How some secondary serials reported an American paper.

Current AIDS Literature also had an Editorial comment, but also had a short comment elsewhere about the other agents

Editorial comment on papers of outstanding interest
PP Clapham, Chester Beatty Laboratories, London, UK

Dissection of the processes occurring during the infective pathway of HIV will be essential to elucidate events which may be potential targets for antiviral agents and to define functional domains of HIV (glyco) proteins which could be the targets of neutralizing antibodies and thus be relevant to vaccines development.

Dextran sulphate, a long chain polymer of glucose, used for over 20 years as an oral anticoagulant, was recently shown to inhibit HIV *in vitro* at concentrations achievable *in vivo* (Ueno and Kuno, *Lancet* (1987) i:1379). Mitsuya *et al.* [2] demonstrate that dextran sulphate acts *in vitro* by preventing gp120 binding to CD4, and thus blocks subsequent syncytium induction or infection. HIV-2 was inhibited as well as HIV-1 and it remains to be seen whether the action of dextran sulphate specifically blocks HIV binding or may block a range of receptor/ligand reactions. Hopefully, dextran sulphate may be the first of many agents to be discovered that will block HIV entry into cells.

2. Mitsuya H, Looney DJ, Kuno S, Ueno R, Wong-Staal F, Broder S (Clinical Oncology Program, Division of Cancer Treatment, National Cancer Institute, Bethesda, MD 20892, USA) Dextran sulfate suppression of viruses in the HIV family: inhibition of virion binding to CD4 cells. *Science* (USA) (1988) 240 (Apr 29) 646-649. **See also Item 1142**.

1142 ●● MITSUYA H, LOONEY D J, KUNO S, UENO R, WONG-STAAL F, BRODER S (Clin. Oncol. Prog., Div. Cancer Treat., Nat. Cancer Inst., Bethesda, MD 20892, USA.) **Dextran sulfate suppression of viruses in the HIV family: inhibition of virion binding to CD4+ cells.** *Science, USA* (1988) 240 (Apr. 29) 646-649

Dextran sulphate (M_r ~ 8000) at ≥1 μmol/l inhibited HIV-1 RTase, inhibited syncytia formation of infected cells and blocked binding of HIV to T lymphocytes (but it also affected mammalian DNA polymerase). It also inhibited HIV-2 and showed synergy with dideoxycytidine and dideoxyadenosine.

The MEDLINE printout occupied a whole page and included the abstract. The 'Descriptors' from the MESH Thesaurus are chosen by the staff at NLM and those with a star show under which headings in *Index Medicus* the paper will be noted. Twenty descriptors are given, but the effect is bewildering. CAS Registry numbers and the Enzyme Commission, EC number are given. The analogs and derivatives and suramin are noted, but not their significance.

FIG. 10.1: continued

```
  Tags: Comparative Study
  Descriptors: *Dextrans--Pharmacology--PD; *Helper Cells--Microbiology--MI
;   *HIV--Drug    Effects--DE;   *Virion--Drug   Effects--DE;   Antigens,
Differentiation,   T   Lymphocyte;  Cell  Line;  DNA,  Viral--Analysis--AN;
Fluorescent  Antibody Technic; Helper Cells--Drug Effects--DE; Helper Cells
--Immunology--IM;   HIV--Genetics--GE;   HIV--Physiology--PH;   Immunologic
Technics;  Retroviridae  Proteins--Physiology--PH;  Reverse  Transcriptase
--Antagonists  and  Inhibitors--AI;  Reverse Transcriptase--Metabolism--ME;
Suramin--Pharmacology--PD;   Thymidine--Analogs   and   Derivatives--AA;
 Thymidine--Pharmacology--PD; Viral Fusion Proteins--Physiology--PH; Virion
--Physiology--PH
  CAS Registry No.: 0   (HIV envelope protein gp120); 145-63-1   (Suramin);
30516-87-1   (zidovudine);  3416-05-5   (2',3'-dideoxythymidine); 50-89-5
 (Thymidine);   9042-14-2   (dextran  sulfate)   Enzyme  No.:  EC  2.7.7.-
 (Reverse Transcriptase)
```

Searching *Biological Abstracts* is tedious and using the KWIC index is not useful in this case:

· CATHETER **CELL**

Subject Context	**▼ Keyword**	**Ref. No.**
UCTION ELISA/ SOLUBLE	**CD23** IS RELEASED BY B LYMPHOCYTES C	15739
PATHWAY/ ANALYSIS OF	**CD3** AND ANTIGEN RECEPTOR EXPRESSIO	15746
I. DECREASE OF CD2 AND	ANTIGEN EXPRESSION HUMAN PLAT	15753
D BETA-CHAINS WITH THE	COMPLEX RABBIT CELL SURFACE EX	15706
EXPRESSION OF LY-2 AND	**CD3-ALPHA-BETA** T CELL RECEPTOR IN C	15749
NCE/ TI-WT31-NEGATIVE	**CD3-POSITIVE** LARGE GRANULAR LYMPH	17051
OF A REQUIREMENT FOR	**CD3-T** CELL RECEPTOR COMPLEX EXPRES	17362
PECIFICITY AFTER MOUSE	**CD4** GENE TRANSFER INTO A T HELPER H	15732
PLEX MOLECULAR MASS/	MOLECULES ARE ASSOCIATED WITH	15688
IBLE MODULATION OF T4	ON MONOCYTOID CELLS BY PHORBO	15851
NDROME/ MOUSE BRAIN	TRANSCRIPTS ENCODE ONLY THE CA	15825
PER HYBRIDOMA MURINE	**CD4-EXPRESSING** TRANSFECTANTS T-CEL	15732
ULATE LYTIC ACTIVITY IN	**CD4-POSITIVE** AND CD8-POSITIVE ALLOR	15684
N OF VIRION BINDING TO	CELLS HUMAN IMMUNODE	12780
AND PROLIFERATION OF	PERIPHERAL BLOOD T LYM	15736
MUNODEFICIENCY VIRUS	T CELL SUBSET DECREASE	16821
ATION OF L3T4-POSITIVE	T LYMPHOCYTES MOUSE I	15752
UNCTIONAL ANALYSIS OF	**CD8** LYMPHOCYTES IN LONG-TERM SURVI	15873
LETION OF LOW-DENSITY	POSITIVE CD16 POSITIVE LYMPHOCY	15760
ITY IN CD4-POSITIVE AND	**CD8-POSITIVE** ALLOREACTIVE CLONES HU	15684
TE-EARLY PROTEIN PP89	T CELLS VACCINE/ A NONS	15389

FIG. 10.1: continued

layout is off-putting and the various descriptors are not very helpful, although searching by CAS Registry number or enzyme number was possible.

The shortcomings of the large secondary serials, and the abstracts written by authors, are clear. On the other hand, the informative, selective newsletters are very efficient and quick to spot and draw attention to important papers. Who will do the same for science and will scientists appreciate and pay for such services?

Use of publications and online databases

Publishers keep their sales figures and customers secret. The Bureau of Hygiene, however, allowed me access to their data on sales and users of *AIDS Newsletter*, *AIDS and Retrovirus Update* and the database (Wyatt, 1988). The use reflects not only the other services available, but also the money available (Tables 10.8 and 10.9). There were several other newsletters produced in the USA, but no comparable database. The relative budgets of the types of users are reflected by the usage: it is difficult to imagine UK schools paying for access to a database on AIDS, whereas this is common in the USA. Part of the use reflects the location of major pharmaceutical companies which might be interested in therapies for HIV infection and AIDS. Although these data are for a specialized service and are now dated, they probably reflect the overall use of services.

Picking the best

The Howard Hughes Medical Institute (HHMI) funds 222 top-quality workers in 53 institutions. HHMI now produces more than 20% of the 200 most-cited biomedical papers. The Great Neglected Diseases of Mankind GND programme of the Rockefeller Foundation funded 14 outstanding research laboratories for eight years: the 1280 publications had a mean impact factor of 3.94. Only 130 of the more than 4000 journals in the *Science Citation Index* had greater impact factors (The Lancet, 1994). If only the very best and productive scientists are generously funded, and many are likely to be young, they will probably continue to be highly productive. Their work is, nevertheless, founded on the work of all the others.

Table 10.7. Availability of an American paper to UK users: arrival at the British Library Document Supply Centre, Boston Spa, UK

Cover date 1988	Arrival date 1988	Serial	Country of origin	
29 April	1 June	*Science*	US	
26 May	27 May	*New Scientist AIDS Monitor*		UK
3 June	30 June	*AIDS Newsletter*	UK	
June	4 July	*ATIN*	US	
	c8 July	online Medline	US	
June	22 July	*Current AIDS Literature*		UK
July	9 August	*AIDS Information*		UK
15 July	22 August	*Biological Abstracts*	US	
	August	*Index Medicus* 1988 – CD-ROM	US	
August	14 September	*AIDS* (bibliography)		UK
September	27 September	*AIDS Forschung* in German		
August	28 September	*Index Medicus*	US	
July	6 October	*AIDS Research Today*	US	

The paper investigated in this table was chosen because it was published in a US journal, was important, recent and covered by all the secondary serials. It was difficult to choose a paper which fulfilled these criteria – more than a dozen papers were tried. I did not find it in *RSM AIDS Letter*. In the *AIDS* bibliography it appeared under 'immunology'. Although US journals are not sent airmail, their arrival at BLDSC is fast: the agency may use bulk airfreight. The October *Scientific American*, which arrived on 20 October, contained an article by Mitsuya *et al.* on this topic but did not include a reference to the *Science* paper.
ATIN is the usual name of *AIDS Targeted Information Newsletter.*

Impact and other factors

Of the 4398 science journals (8%) in the *Journal Citation Index* only 347 garnered 54% of all citations. Library studies show that 10% of journal titles account for 80% of use. Only 45% of all papers in the *Science Citation Index* 1981–1985 were cited once or more times. In one study a complex Medline search provided 1000 citations, of which only 150 were relevant, and only three of these were found to be methodologically sound.

The impact factor and journal rankings

The impact factor (IF) is calculated by taking the number of all citations for a particular journal for the two previous years, and dividing this by the total number of articles published in the journal in this time. Only articles that cite within two years after the publication of the article, contribute to the IF. This assumes that citing occurs most frequently in these two years, which may

Table 10.8. Use of the services of the Bureau of Hygiene and Tropical Diseases: percentage by type of user

User	UK distribution only		World use of database
	AIDS Newsletter	*AIDS and Retrovirus Update*	
Doctors, hospitals, medical services	26	32	30
Health administration	11	7	0
Education, counselling	33	20	0
Medical schools, universities, research	5	13	33
Professional, government, international bodies	8	4	3
Pharmaceutical etc. companies	6	7	18
Newspapers, TV etc.	5	4	0
Other	7	4	18*

* including 6% by community users, e.g. public libraries, schools and life insurance firms
AIDS Newsletter and *AIDS and Retrovirus Update*: figures for 1987 Database users were those using the database in November 1987. *AIDS and Retrovirus Update* is now named *Current AIDS Literature*

be true of 'here today and gone tomorrow' disciplines, but this does not apply in ecology and systematics. Journals are ranked by their IF: the higher the IF, the more important a journal is considered to be. Journal ranking and IF are now used to judge the research effectiveness of individual scientists, departments, institutions and even nations, but this has been criticized by a number of scientists (Hansson, 1995). The ranking of a journal with a high IF has no bearing on the citing of particular articles: some may not be cited and the ranking may be determined by many citations to a few papers. See also p. 166.

Where scientists wish to publish may not be determined by the IF of the journals (Table 10.10). Scientists wish to influence their peers and to communicate their findings to scientists who will read the journals in which they are published.

Review serials

One feature of the *Journal Citation Reports* lists is the very high ranking of the annual review-type serials in some subjects and their absence from others. Some subjects have very high IF, whereas others are very low (Table 10.11). Most of these review serials are annuals but some, e.g. *Microbiology Review*, *FEMS Microbiology Review* and *Quarterly Review of Biology* appear in parts. What are the reasons for these very wide subject differ-

Table 10.9. Users of the Bureau of Hygiene and Tropical Diseases services: percentage by country

	AIDS Newsletter	*AIDS and Retrovirus Update**	Database
UK	78	64	17
West Germany	2	2	4
Switzerland	1	2	4
Other European countries	8	14	12
North America	4	5	61
Africa	2	8	0
Middle East & Mediterranean	1	0	}
Far East	1	1	} 2
Australia & New Zealand	1	3	}

* now *Current AIDS Literature*

AIDS Newsletter and *AIDS and Retrovirus Update* (now *Current AIDS Literature*): figured for 1987

Database users were those using the database in November 1987

ences? Why do biotechnology and the agricultural sciences have such low IF, even for review serials?

Comparing subjects

There are changes from year to year (see Table 18.3). Thus the number of journals for biochemistry and molecular biology was 151 in 1991; by 1993 this had risen to 173 and the IF for *Annual Review of Biochemistry* had dropped from 48.3 to 37.9, although it was still first. For genetics and heredity, *Annual Review of Genetics* dropped from first to third place and its IF dropped from 15.1 to 12. For immunology, *Annual Review of Immunology* jumped from an IF of 25.4 to one of 37. At the Eighth International Conference of the International Federation of Science Editors in July 1995, Eugene Garfield, commenting on the use of IFs, suggested that different types of articles, e.g. reviews, research articles and letters, should be analysed separately. Comparison between subjects or fields was not valid: for purposes of evaluation, even fields might be subdivided. Garfield compared the ranking of the top 34 research journals on their 1993 IF against the journals' cumulative IF (all citations 1981–1993) for articles published in 1981, and their rank according to their cumulative factor: the two rankings were strikingly different. He also pointed out that research productivity should not be based on IF, as IFs

Table 10.10. Choice of journals by 50 intramural NIH scientists (from Foster, 1995)

Journal	NIH scientists' rank	ISI rank
Science	1	10
Nature	2	8
Cell	3	3
PNAS	4	37
J Biol Chem	5	67
J Cell Biol	8	35
Biochemistry	9	111

do not distinguish between favourable and unfavourable citations (see p. 25) (Garfield, 1996).

The retiring editor of *Proceedings of the Society for Experimental Biology and Medicine*, which published many short papers had kept a record of the the work of its 37 editors (Goldfarb, 1961). Newly appointed editors were either too severe (up to 91% rejection) or over-lenient, although all tended to conform to their subject group when standards were discussed. For different subjects there were rejection rates of 7–24% in one and 7–53% in another, rates which seemed constant and based on subject norms. Other criteria also showed constant, wide variations in standards.

Two recent products from the ISI database are: *National Science Indicators on Diskette 1981–1994*, a dataset of research impact indicators based on citation data recorded for 17 disciplines and 200 countries. Biomedical disciplines include molecular biology and genetics, biology and biochemistry, neurosciences and immunology (see p. 00). *Index of Expected Citations for Science Papers, 1981–1994*, shows the number of citations attracted by papers in any of 3500 scientific journals. The 89 subject headings do not correspond to those in the *Journal Citations Reports*. There will be few copies of these products available, but searches may be made at the Wellcome Centre for Medical Science, London NW1 2BE, UK. A typical suggested query is 'should I publish my article in *Annals* of *Internal Medicine* or *BMJ* to get the highest impact?' This seems to me to be the wrong question: any citation will be to the article, not the journal, and the first question is whether the article is likely to appeal to the readership of the journal and hence be considered by the editor. For some disciplines, especially clinical ones, the IF may not be applicable.

Bias in the selection of the literature

There is always a tendency to use one's own language because it is easier. Thus for many years complement research languished until a generation of German speakers went back to the original papers of Erhlich. European scientists often feel that their work is not given enough credit or citations by Americans. When important books or series, such as the *Annual Reviews in*. . . ., are published in a particular country, it is inevitable that there may be a bias (Table 10.12), which can be reinforced by limited exposure. Of journals recommended for library holdings for undergraduate biology in the US, 81% were from the US, 17% from the UK and only 2% from the rest of the world (Wyatt, 1972b).

BIDS

In the UK the Bath Information and Data Services, BIDS, has negotiated licences from databases on behalf of academic institutions. Each institution pays an annual fee for the services it requires, and the use of the service is then free for members of staff and students. The first large database was the ISI, comprising the Citation Indexes, and ISTP, which effectively includes the *Current Contents* series. This service has been very popular, as most institutions subscribe and it can be accessed from personal terminals. Search strategies can be saved and reused and results may be downloaded locally. One need never go to the library, as reprints may be ordered by e-mail.

Since the start in 1991, BIDS has grown and the following major services are available: ISI; EMBASE, the *Excerpta Medica* database; Inside information, the British Library current awareness service; UnCover, the current awareness service from Blackwells; the CAB Public Health and Tropical Medicine database; Ecological Flora of the British Isles; the Checklist of the British Flora; *Kew Record*; and, *LegumeLine* (p. 242).

Services such as EMBASE are likely to be used only by institutions with medical faculties. In the same way the botanical databases will be used by only a few institutions. More services will be added in the future.

Table 10.11. Citations of review journals by subject classification

Subject	No. of journals	Rank	Review journal	Impact factor
Biochemistry and molecular biology	151	1	*Annu Rev Biochem*	48.3
		2	*Annu Rev Cell Biol*	24.2
		4	*Adv Cyclic Nucl Prot*	18.2
		5	*Adv Prot Chem*	16.5
Immunology	80	1	*Ann Rev Immunol*	25.4
		2	*Adv Immunol*	16.4
		5	*Immunol Rev*	9.7
Microbiology	69	1	*Microbiol Rev*	16.3
		2	*Adv Microbiol Ecol*	8.8
		3	*Ann Rev Microbiol*	8.4
		4	*FEMS Microbiol Rev*	4.1
Genetics & heredity	61	1	*Ann Rev Genet*	15.1
		2	*Adv Hum Gen*	6.7
Botany	121	1	*Ann Rev Plant Physiol*	13.4
		2	*Ann Rev Phytopathol*	4.5
		4	*Adv Bot Res*	4.1
Ecology	45	1	*Adv Microb Ecol*	8.8
		2	*Ecol Monogr*	5.3
		3	*Adv Ecol Res*	4.7
		4	*Ann Rev Ecol Syst*	3.6
Virology	18	2	*Adv Virus Res*	4.6
		5	*Prog Med Virol*	2.2
Biology	50	2	*Q Rev Biol*	3.8
		3	*Biol Rev*	3.8
Biotechnology/applied microbiology	30	4	*Adv Appl Microbiol*	2.1
Agriculture	103	8	*Adv Agron*	0.7
Agriculture, dairy animal science	28	23	*World Anim Rev*	0.1
Zoology	96	—	—	—

Prepare from the *Journal Citation Reports* 1991

Titles and summaries

One problem is that papers titles were originally meant to be read in conjunction with the printed journal. The information is a compromise between the arresting 'either or' (about fungal taxonomy) and the long titles full of acronyms, designed only for specialists. Keywords added to the abstract are unhelpful when you are reading only the title. If each paper were assigned a unique identifying number (p. 26), it would be possible to devise more informative titles and descriptions to suit the variety of ways in which we can search for information.

Journals should publish more articles of interest to general readers rather than specialists. More journals such as *Biologist* should have items like the Minerva page of the *BMJ*, a series of short summaries of interesting work culled from a wide reading of the medical literature.

Literature spread

The BMA Guide to Rabies, edited by D.R. Morgan, (Radcliffe Medical Press, 1995) examines the pros and cons of vaccination of foxes and other animals against rabies, and the policy of quarantine. The book is an example of the complexity of a problem which is political, but based on medical opinions and data from zoological and veterinary sources. Of 137 references, one is to *Nature* and two are personal communications from persons in government and industry. Twenty-three medical journals are cited, but several are obscure and would not appear in the ISI database. Six veterinary, seven zoological and one food journal are cited, spanning several other databases. Seven references to newspapers and other general sources, including *New Scientist* and *Trends in Microbiology*, might not appear in any database. There are extensive citations from eight books, two annuals and four symposia, which might be difficult to find through the databases and secondary services. Six references to international agencies and several of the items are not published in the ordinary sense. There are ten British government (HMSO) publications and another three from British agencies, the Nature Conservancy, the Joint Nature Conservation Commission and the Department of Trade and Industry, which would be picked up in Sigle. There are three other documents from Europe and one from Australia, as well as an unpublished EEC document. Other items cited are an unpublished BA thesis, the *Law Society Gazette*, the Pet Food Manufacturers' Association, UFAW and the *Shooting Times and Country Magazine*. Such is the complexity of real information needs in a multidisciplinary publication. Such a variety of sources comes from the accumulated experience of a group of scientists and their friends and colleagues.

Table 10.12. Percentage of serials and citations originating from the USA or North America*

Subject	Serials	Citations	From
Zoology	20*		*Zoological Record* 1994
Biology, biomedical		42	*SCI* 1979
Genetics, biophysics	47	57	*Annual Reviews*†
Biology		28*	BIOSIS database 1990

†From Sengupta (1984)

The spread of subjects

Although most scientists will know something of the structure of the literature of their subject from looking at journals, very few will have taken the time to make a quantitative study of some of its features. Some subjects have a very compact core of journals, whereas others have papers scattered through many journals. Vaccines is not a separate subject in the *Journal Citation Reports*, but although there are few papers in any journal, half of the papers were published in just ten journals (See Table 10.3). Since this study there has been an increased interest in vaccines and a new specialist journal, *Vaccine*.

Medical bacteriology is a very large subject spread over hundreds of journals, and only about one-seventh of the papers are in the top 10% of journals. A worker on vaccines would need to scan very few journals, whereas a medical bacteriologist would have to use a secondary serial, searching only for the organisms or terms of interest. The *Journal Citation Reports* has no division for medical microbiology, but the microbiology section lists only 69 journals and few are specifically medical. If longer periods were scanned the number of journals would increase, but the number of new significant papers, as defined by *SCI*, would be very small. Nevertheless, the number of journals cited in only six months by *Microbiological Abstracts* for medical bacteriology alone, was 521.

Tables 10.3, 10.4 and 10.6–10.9 and Figure 10.1 are reproduced from Wyatt (1988) by kind permission of The British Library.

References

Ayres J.E., Taylor A.E.R. Online searching of parasitology literature. *Parasitology Today* 1988; 4: 324–328.

Carson J. *The Microbiologist and his Literature.* PhD thesis, University of Bradford, UK, 1984.

Davison P.S., McDonagh M., Meldrum J.D.P., Moss A. *International bibliographic*

review on costs and modelling in information retrieval. British Library Research Paper 37, 1985.
Foster W.R. Impact factor as the best operational measure of medical journals. *Lancet* 1995; 346: 1301.
Garfield E. How can impact factors be improved. *British Medical Journal* 1996; 313: 411–413.
Goldforb A.J. A brief history of the Society for Experimental Biology and Medicine. *Proceedings of the Society for Experimental Biology and Medicine* 1961; 107: 702–712.
Hansson S. Impact factors as a misleading tool in evaluation of medical journals. *Lancet* 1995; 346: 906. See also letters *Nature* 1995; 376: 720.
Sengupta I.N. A weightage formula to rerank biological publications. *International Library Review* 1984; 16: 447–459.
The Lancet. Proposal quality or product quality? *Lancet* 1994; 344: 488–489.
Wyatt H.V. When does information become knowledge? *Nature* 1972; 235: 86–89.
Wyatt H.V. More foreign journals. *BioScience* 1972; 22: 570.
Wyatt H.V. How history has blended. *Nature* 1974; 249: 803–805.
Wyatt H.V. *AIDS Information: Supply and Demand*. British Library Research Paper 55, 1988.

CHAPTER ELEVEN

Major secondary services including CD-ROMs

H.V. WYATT

Now that electronic bibliographic databases are commonplace, it is possible to produce different versions in a variety of formats, and to package subsets for particular markets and interests – at a price. Some packages are available in other forms, whereas a few are currently only available as CD-ROMs. Material is repackaged to conform to sales, so that the titles and contents of the different sections of some of the databases are changing without much notice, and so may be different from those listed here. The printed versions may still be preferred by some, and indeed have some advantages, particularly as they can be photocopied and perused at leisure. In general, an indexing serial will be more timely than one which contains abstracts, and both will be more up to date than one with evaluated abstracts. *Biological Abstracts* obtains details from some journals, mainly from the USA, while the papers are still in press. For *Index Medicus* some foreign papers are processed by collaborating centres abroad, e.g. in Britain and Sweden.

Some serials have a very limited print run and are therefore to be found in very few libraries. There are about 100 universities in the UK and over 3000 colleges in the USA. Although libraries have cut their subscriptions to avoid duplication, there are still many colleges with more than one campus and library. There are research and government institutions and commercial laboratories. A print run of just over 100 copies eg *Immunology Abstracts* means that only very large centres will have access to a serial except by online searches or CD-ROM.

Specialized services are included in the subject chapters, e.g. biotechnology.

Table 11.1. The BIOSIS services

Service	Frequency	No. of items	Print	CD	Online	Microfilm
Biological Abstracts	two weeks	370 000	+	+	+	+
Biological Abstracts/RRM	monthly	260 000	+	+	+	–
Zoological Record	two weeks	70 000	+	+	+	(+)
Abstracts of Entomology	monthly	20 000	+	–	–	–
Abstracts of Mycology	monthly	20 000	+	–	–	–
BIOSIS GenRef	quarterly	100 000	–	+	–	–
BioBusiness	weekly	90 000	–	–	+	–
Foods Intelligence	quarterly	50 000	–	+	–	–
For these last two services see p. 136						

Biological

BIOSIS

The major biological secondary service is provided by BIOSIS as three publications, *Biological Abstracts (BA), Biological Abstracts/ RRM* (for reports, reviews and meetings) and *Zoological Record. Abstracts of Entomology* and *Abstracts of Mycology* are also produced monthly from the main database, with cumulated annual indexes. The database is compiled from 9000 serials, books etc. and since 1986 from certain classes of US patents: it contains about 10 million references from 1969. *Biological Abstracts* is based on about 7000 journals, whose contents are allocated to one of 84 subjects: new taxons are identified by an asterisk. *Biological Abstracts/RRM* is compiled from meetings, conferences, workshops and symposia, review articles, books and software. In 1990, Europe and the Middle East accounted for 48% of the items, 28% were from North America, 16% from Asia and Australia, 6% from South and Central America and 2% from Africa.

The two services *BA* and *BA/RRM* comprise the online database BIOSIS Previews, which contains about 10 million items. Six-month cumulated indexes are made for *BA* and an annual one for *BA/RRM*, but the items themselves are not cumulated: searching of the printed volumes is thus very laborious and frustrating and is not recommended. Searching online or of CD-ROM is to be preferred. For *Zoological Record* see p. 203. (For more information on zoology, see Chapter 22).

The BIOSIS Information Transfer System or BITS is a monthly literature search diskette based on the researcher's needs from the BIOSIS database.

BIOSIS GenRef is a new service with international gene numbers added whenever they appear in the article.

Table 11.2. The CABS serials

Current Advances in . . .	No. sections	No. titles p.a.
Applied Microbiology & Biotechnology	182	12 000
Cancer Research	211	15 000
Cell & Developmental Biology	197	21 000
Clinical Chemistry	175	7 000
*Ecological & Environmental Sciences**	177	22 000
Endocrinology & Metabolism	223	13 000
Genetics & Molecular Biology	172	27 000
*Immunology & Infectious Diseases**	302	24 000
*Neuroscience**	277	22 000
*Plant Science**	218	23 000
Protein Biochemistry	174	19 000
Toxicology	154	8 500

* available on floppy disk

The taxonomic reference file (TRF) of BIOSIS contains 13 000 bacterial names, including synonyms and type strain data, with 12 000 citations with full abstracts from 1969. Also included is access to LIMB (see p. 142).

BIOLOGICAL ABSTRACTS, BA/RRM, BIOSIS GenRef and FOODS INTELLIGENCE are the names of the CD-ROMs. The number of items added varies from year to year in the literature about these services.

Current Awareness in Biological Sciences (CABS)

This serial has a long history under different names from 1919, and became known in its present form in 1983. It should not be confused with CABI, below. It is published monthly in 12 subject sections, whose titles have often changed to reflect new interests. Some 2000 journals are scanned. For some journals all papers are entered, but for others only about 75% of the articles are deemed relevant. Some have only a few papers selected. The journals are chosen for relevance, but the articles in the journals must be predominantly in English. The weeklies and priority journals are entered as soon as they arrive. Each entry is numbered and consists of title, author(s), bibliographic details and address. At the end of each section 'other relevant titles' are listed. A paper may appear in more than one section. The database may be searched online for title, journal etc., and by species and keyword field. Several of the sections are available on floppy disk and more sections will be added. One feature is that the section headings are changed with new research interests. The archive database contains more than 2 million records, and about 160 000 items are added each year.

The Life Sciences Collection (LSC)

There are 19 subject subfiles from the Cambridge Scientific Abstracts database (CSA) covering 5000 journals, books etc. and adding about 130 000 abstracts each year. The entire database from 1982 may be searched online, an magnetic tape and an CD-ROM. LIFE SCIENCES COLLECTION 1982 – is the complete CD-ROM collection of the CSA serials, with 1 500 000 items. However, CSA also publish *Biotechnology Research Abstracts*, also called Cambridge Biotechnology Research Series, and prior to 1993 known as *Agricultural and Environmental Biotechnology*. At least 14 other information providers, including EMBASE and MEDLARS, may also be accessed through the CSA WWW site and, to make it more complicated, CSA files may be accessed through other providers as well. On the WWW the files are updated monthly and there is a five-year backfile. The LSC serials are listed in Table 11.3 p. 80.

In addition, CSA publishes *Aquatic Sciences and Fisheries Abstracts, ASFA* 1971– in three parts: *ASFA Part 1 Biological Sciences and Living Resources*, with 22 000 items; *ASFA Part 2 Ocean Technology, Policy and Non-Living Resources*, with 9000 items and *ASFA Part 3 Aquatic Pollution and Environmental Quality*, with 7000 items a year. This series is compiled with the cooperation of FAO and others from over 5000 serials, books, conference proceedings, translations and limited distribution literature.

ASFA Aquaculture Abstracts 1984– is a subset of the ASFA-database with 5000 abstracts a year. *Ocean Abstracts* 1964– has 12 000 items a year.

There is also an Environmental RouteNet on the WWW, updated daily, with access to a very wide variety of information and organized in 23 subject areas, e.g. agriculture including pesticides and market news; molecular biology databases and conservation including biodiversity and endangered species. (URL: http://www.csa.com/routenet/ for the home page.)

The Science Citation Index, SCI 1955–

This vast database is derived from the journals used for *Current Contents*. It is assumed that authors will cite key papers and that a subject may be followed through these citations, i.e. papers in which PCR is used will cite a key paper about PCR. However, this may not be the case, as some authors may consider a paper to be so well known or obvious that it need not be cited, or key papers may be deleted for lack of space. Also citation of a paper may be for less obvious reasons: several citations to Avery's paper concerned the use of an enzyme, DNAase, not the transformation by DNA (Wyatt, 1972). One good use of *SCI* is to look for modifications of a technique, as the original method will always be cited. The most frequent use of *SCI* is to see how many times one's own papers have been cited. The database may be searched to provide a current awareness service e.g. via BIDS (p. 69).

One can search for a particular author, for citations to a particular paper, for institutions, and, using the Permuterm index, for significant words from the titles of papers. The Source Index lists organizations by geographical location and institution, and also by name, as well as giving the addresses of all source authors who have cited papers.

The database also has the *Index to Scientific Reviews* 1974–, *Index to Scientific Book Contents* 1985–, and is used to produce the *SCI Journal Citation Reports* 1975–. The journal impact factor (IF) is now widely used for evaluation of researchers, departments and institutions, as well as the status of the journals themselves. Although the IF gives important information about journals within the same subject, comparisons between subjects may be very difficult and misleading (see p. 65).

PASCAL

The French abstracting serial *Le Bulletin Signalétique du CRNS*, in which the abstracts were in French, was replaced in 1984 by PASCAL *Bibliographie Internationale*, now available in print, online, on disk and on CD-ROM. Most abstracts are now in English. About half a million items are added each year from 9000 journals, 6000 theses, 3500 reports (mainly French). 2000 congresses and 1000 other items. About 63% of the items are in English, 12% French, 10% Russian, 8% German and 7% other. The old files 1973–1984 are available and hold 6 million items. PASCAL covers all the sciences and consists of two databases. *Pascal Sigma* is multidisciplinary, with 300 000 items per year, of which S2 is biology and S3 is medicine. *Pascal Thema* has the more specialized subjects, such as T215 biotechnology, T260 zoology, T280 agriculture and T235 tropical medicine. *Pascal Folio* contains groups from both *Sigma* and *Thema*. Similarly, *Pascal Explore* contains some medical sections from both *Sigma* and *Thema* which were not in the *Bulletin Signalétique*. There is a table of correspondence for the different parts of the two serials.

Table 11.3. The LSC serials

Title			Number of abstracts	
			Per year	In subfile*
Animal Behavior Abstracts	1972–		5 200	85 000
Cambridge Scientific Biochemistry Abstracts				
Amino Acids, Peptides & Proteins Abstracts		–1993		134 000
Biological Membranes Abstracts		–1993		127 000
Nucleic Acids Abstracts	1970–		14 000	151 000
Cambridge Biotechnology Research Abstracts				
Agricultural & Environmental Biotechnology	1993–		3 600	4 600
ASFA Marine Biotechnology	1989–		1 600	8 000
Medical & Pharmaceutical Biotechnology	1993–		3 600	5 000
*BioEngineering Abstracts**	1974–		11 000	
Calcium & Calcified Tissue Abstracts	1969–		3 200	53 000
Chemoreception Abstracts	1972–		1 500	26 000
CSA Neurosciences Abstracts	1982–		16 500	180 000
Ecology Abstracts	1974–		13 000	190 000
Entomology Abstracts	1969–		10 500	162 000
Genetics Abstracts	1968–		15 400	220 000
Human Genome Abstracts	1990–		3 600	19 000
Immunology Abstracts	1976–		13 000	222 000
Cambridge Microbiology Abstracts Series				
A: Industrial & Applied Microbiology	1965–		7 700	153 000
B: Bacteriology	1966–		13 000	235 000
C: Algology, Mycology & Protozoology	1972–		10 000	175 000
Oceanic Abstracts	1964–		12 000	
Oncogenes and Growth Factor Abstracts	1989–		2 000	15 000
Pollution Abstracts	1970–		11 000	
Toxicology Abstracts	1978–		9 000	145 000
Virology and AIDS Abstracts	1967–		11 000	140 000

*Not part of the Life Sciences Collection

Medical

Index Medicus and MEDLARS

This major medical service also covers many subjects of biological interest: parasitology, immunology, microbiology, nutrition and

Table 11.4. How *Index Medicus* has coped: subheadings and entries (from Wyatt, 1988)

Heading	1980	1981	1982	1983	1984	1985	1986	1987	1988
Retrovirus	+	+	+	+	+	+	+	+	+
HTLV					+	+	+	+	+
Subheadings					12	13	14	19	
Entries					200	525	850	190	550
HTLV III now HIV								+	+
Subheadings								13	
Entries								750	1400
AIDS				+	+	+	+	+	+
Subheadings				17	22	29	30	30	
Entries				310	740	1250	1700	2250	3000
AIDS R C								+	+
Subheadings								22	
Entries								200	320
Total no. of entries				310	940	1775	2550	3400	5300

The number of entries was estimated from *Cumulated Index Medicus*. The numbers for 1988 are extrapolations from the number of entries January–June 1988. Most citations will have more than one entry.

many others. Many (almost) pure life science journals are extensively used, and *Index Medicus* should be seriously considered as the service of choice. Unlike *Biological Abstracts*, only journals (about 2700) are considered and abstracts are only available in the online and CD-ROM versions. The monthly parts are cumulated annually and are therefore much easier to search than *BA*. All papers are placed in subjects according to a limited vocabulary given in the annual *Medical Subject Headings (MESH)*. It is very important that *MESH* is consulted for each year of the search and the right subject headings noted. Table 11.4 shows how *Index Medicus* has coped with more subject headings and subheadings as the number of papers has increased. One difficulty is that until a subject has been in existence for some time and its importance recognized, there will be no proper subject heading for it. Thus early papers relating to the virus causing AIDS were difficult to find (see also p. 60). The number of papers relating to AIDS and HIV in *Biological Abstracts* in the early years is, in fact, inflated, because all papers on HTLV I and II were included. Similarly, papers on thalidomide were not given a separate subject heading in *Index Medicus* until after the crisis had passed.

The MEDLARS database at the Karolinska Institutets Bibliotek och informationscentral in Stockholm operates an extensive loan and photocopying service as well as being an online host. It has two associated databases. SWEMED 1982 – contains more than 10 000 references to biomedical journal articles and reports published in Sweden and all medical dissertations at Swedish universities. NORDSER locates biomedical journals and numbered congresses in about 400 Nordic libraries.

Excerpta Medica and EMBASE

This service is based on 41 printed abstract journals, of which the following will be of interest to biologists: *Clinical and Experimental Biochemistry*; *Developmental Biology and Teratology*; *Endocrinology*; *Human Genetics*; *Immunology, Serology and Transplantation*; *Microbiology: Bacteriology, Mycology, Parasitology and Virology*.

About 3500 journals from 110 countries are scanned, but not all papers will be chosen and 60% of those chosen have abstracts. A problem with using the printed versions is that a paper on, say a tumour virus, may appear in an abstract serial such as *Cancer*, rather than the virology serial. Many journals from Europe and Asia are included, although they are not found in other biomedical databases. Although only about 36% of the journals are common to *Excerpta Medica* and *Index Medicus*, the number of papers common to both may be much greater as different selections may be made from minor journals. EMBASE is the database, which may be searched online, via diskette, tape etc. or by CD-ROM. EMBASE 1980 – has more than 3.5 million records and 350 000 are added each year. Files are available from 1974 – for online use. This database is particularly useful for drug information, and about 40% of the records have drug-related information. Compounds are indexed by CAS Registry number and by brand and manufacturer's names. Like *Index Medicus*, the database has to be searched using a limited thesaurus called EMTREE.

As well as drug information, this service emphasizes 'fast-breaking developments, including AIDS, biotechnology and environmental issues'. EMBASE CD: IMMUNOLOGY AND AIDS is a rolling 10-year collection of about 520 000 records. EMBASE ALERT is a rolling 8-week online or CD file. Elsevier, publishers of *Excerpta Medica*, also print about 1000 books each year as well as 1000 journals. EVENTLINE is a database of all kinds of meetings, trade fairs etc. compiled from the resources of the publishers and a network of correspondents. It is an online service.

Food and Nutrition

Food Science and Technology Abstracts (FSTA)

This service, 1969–, is monthly and contains about 18 000 references each year from 1800 journals as well as patents, books, legislation and standards and conference proceedings. The material is gathered from a very wide range of journals, including *European Food Law Review*, reports etc. from all over the world, including Japan, China, the Far East, India, Eastern Europe etc. There is a biotechnology section. *FSTA* should be consulted if there is a query which might involve hard-to-find sources in the range of the subjects covered. There are 18 sections with 130 subsections, and a cumulated annual author and subject index. Material is translated from 40 different and exotic languages. The database may be searched online, by magnetic tape or CD-ROM, and contains 500 000 items. Photocopies of articles from the Far East may be obtained from documents held in the Universiti Pertanian Malaysia. There are 30 'Food science profiles' 1993– in print or on diskette, produced monthly with an annual cumulation, e.g. aflotoxins, pesticides in food. *Viticulture and Enology Abstracts (VITIS)* is also published by the International Food Information Service (IFIS). The *Directory of Research and Education in Food Science, Technology and Engineering* 1993 (2 volumes, one each for Europe and North America) gives details of more than 500 institutions.

Research Trends in Food Science and Technology

This monthly service, started in 1995, indexes about 2000 articles a year and might be of use to some microbiologists and applied biologists. There are also *Research Trends in Pest Management* and *Research Trends in Food Safety. International Food Safety News*, monthly, 1992– , has a few abstracts, articles and news of meetings. All are produced by Research Information of Hemel Hempstead, HP2 7TD, UK.

There are three other CD-ROMs which are relevant: FOOD AND HUMAN NUTRITION 1975 – is also from FAO, with 350 000 records, many unique to this database. *Agrindex* is the print version. HUMAN NUTRITION 1982– is a good example of the best use of combining several databases: CABI, FSTA, Cambridge Scientific Abstracts and NLM. There are more than 550 000 records and 40 000 more are added each year. Between 50 and 70% are derived from MEDLINE. Specific searches of the component databases may yield new references as the combined database is based on automated searches. There is an online human nutrition database, HUNI 1984– , based on several of the CABI abstracting serial. FOODS INTELLIGENCE is produced by BIOSIS (see p. 76) and is derived from over 7600 publications from 90 countries. An archival disc contains 270 000 items, about half of which have abstracts.

Agriculture

The CABI database

The CABI ABSTRACTS database is formed by scanning about 11 500 serials and 4000 non-serial items from over 100 countries. About 700 are core serials, from which most items will be used, and selections are made from the rest. About 85% of the references are from refereed journals. The serials were originally intended for workers in the then British colonies, who had little access to the literature; only relevant items were chosen and the bare bibliographic entry, the abstract or the fuller entry with comments signalled the importance of the item. Workers overseas could obtain a copy of the original through the Bureaux. These serials are therefore very different from the other indexing and abstracting serials. As every item is carefully considered by a named reviewer, the delay in publication time is more than offset by the quality achieved. Because of the wide range of sources used the overlap with other serials is less than might be expected: sources from Europe, Africa and Asia are especially good, American sources less so. The serials range from those of particular interest to life scientists, e.g. parasitology, nutrition, to those of very marginal interest such as tourism.

CD-ROMs are an ideal form of gathering information for sale or gift, to developing countries especially, and several collections are being made. PlantGeneCD covers plant genetic resources, plant breeding and plant biotechnology. Other discs cover animal production, horticulture, crop protection, forestry etc. (see p. 85). Patents are included where applicable. The online and CD data bases have been improved, with new searching terms etc. from 1973–. CABI also produces bibliographies, keys etc., as well as more than 300 books. There are 1500 annotated bibliographies as well as several annuals, e.g. *Seed Pathology and Microbiology*. There are other databases, such as FABIS for the fava bean, which may be searched.

The CABI database should be used in conjunction with AGRICOLA, which has a strong American bias and very little overlap between the two. AGRICOLA is always more recent because it has very few abstracts (see p. 86).

The more biological serials are: *Review of Medical and Veterinary Entomology*, 1913– *Review of Medical and Veterinary Mycology*, 1943– *Review of Aromatic and Medicinal Plants* 1995–, *Nutrition Abstract and Reviews, Series A* 1977– *Helminthological Abstracts*, 1932– *Protozoological Abstracts*, 1977– and *Abstracts on Hygiene and Communicable Diseases* 1926– which includes *Public Health News*. This last contains comments and news from the major newspapers as well as journals and government sources etc. *AIDS Newsletter* 1986– monthly is again an excellent current awareness serial. *Tropical Diseases Bulletin* 1908–.

All these monthly serials have good subject, geographical and author indexes in each issue, with annual cumulations. There are several other more agricultural serials and CD-ROMs (see p. 250) PARA is an online database of medical parasitology and mycology 1984–.

VetCD contains physiology and biochemistry of domestic animals, all aspects of parasites of wild and domestic animals, as well as fungal diseases. It contains about 600 000 records, with 20 000 added each year.

There are several subfile CD-ROMs in addition to the complete CABCD 1984–: BEASTCD 1973–, with 500 000 records about animals, and VETCD 1973–, with 600 000 records on veterinary topics, including parasites. CABPESTCD 1973– has 500 000 informative abstracts about crop protection, nematology, weeds, entomology, pesticides, plant genetics etc. (see also PEST-BANK below); CROPCD 1973– has more than 500 000 informative abstracts, adding 20 000 each year about all kinds of crops; HORTCD 1973– has 500 000 records; PLANTGENECD 1973– has 450 000 records, adding 20 000 more each year; TREECD 1939 – has nearly 400 000 abstracts on all aspects of forestry.

Abstracts on Tropical Agriculture, TROPAG

This service, 1975–, is compiled by the Royal Tropical Institute (KIT) in Amsterdam, with CIRAD in France and NRI, UK and is in print, online and on CD-ROM. TROPAG covers food and industrial crops, agriculture, animals, fisheries, forestry etc. in (sub) tropical regions from 4000 journals as well as books, reports, theses etc. and extension materials. It is monthly, adding about 6000 items each year, with an annual cumulative subject, author and geographical indexes. *Abstracts on Rural Development* (RURAL) is of less interest to biologists and ceased in 1995. Although computer-selected bibliographies can be compiled from

1975, the files 1916–1974 contain about 500 000 titles and can be used for retrospective bibliographies.

Other agriculture databases

AGRICOLA 1970–, from the NAL, with 3.5 million items, includes patents, theses, monographs etc., and covers all aspects of agriculture. WLAS is a list of agriculture serials. AGRINTER is a subfile of the regional agriculture information system for Latin America and the Caribbean. AGRISEARCH has a 10-year cover, with about 75 000 items from five databases in the USA, Canada, Australia, EU, and the Special Program for African Agricultural Research. Subfiles are CRIS (Current Research Information System) for the USA, with about 40 000 items for the current 10 years of projects in agriculture, and INVENTORY OF CANADIAN AGRI-FOOD RESEARCH (ICAR).

Another CD-ROM, AGRICULTURAL RESEARCH PROJECTS (AGREP), lists 20 000 projects of the EU.

The FAO library produces AGRIS with the cooperation of 180 international and national centres, and covers all aspects of plants, agriculture, fisheries, nutrition, pollution etc. There are more than 2.3 million records, including unique material such as unpublished reports, theses, conference reports etc. and more than 130 000 are added each year. Keywords are in English, French and Spanish. Journal articles comprise 75% of records, monographs 18% and conference papers 6%. Nearly a quarter of the references include abstracts. About 16% of the database is non-conventional literature which is not commercially available. *Agrindex* is the monthly bibliography in English, French and Spanish versions. Magnetic tapes are available and the database may be searched online. AGRIS 1975– is the CD-ROM, which has several discs. There are three subfiles available as separate discs: AGRIS FHN 1975– has about 350 000 references to food and human nutrition; AGRIS FIA 1975– has about 125 000 references to fisheries and aquaculture; AGRIS FOR 1975– has about 165 000 references to forestry and agroforestry. See also p. 97.

Environmental

ECODISC 1990– is a version of *Ecological Abstracts*, with 100 000 abstracts and 10 000 added per year. ENVIRONMENT QUALITY (E-CD) 1973–, another CABI subset, covers socioeconomic aspects of agriculture, forestry etc. OCLC ENVIRONMENT LIBRARY 1980– has 700 000 records and adds 25 000 each year. ENVIRONMENTAL

MANAGEMENT 1984– is a CD-ROM with 700 000 abstracts from 7000 international sources, and includes the US federal environment impact statements (EISs). ENVIRONMENTAL IMPACT STATEMENTS 1985– contains abstracts of all US federal government EISs, with 4000 records and 500 more each year. AQUATIC SCIENCES and FISHERIES ABSTRACTS (ASFA) 1978– has 500 000 records (see p. 78).

In addition there are several CD-ROMs which are collections from several different databases: PEST-BANK is a combination of two databases with 27 000 currently registered US pesticides and 40 000 cancelled products. There are names and synonyms, companies, formulations, permissible residue levels etc. POLTOX, for pollution and toxicology, has three separate entities: POLTOX I: NLM, CSA, IFIS 1966– is formed from seven databases: the TOXLINE subfile of NLM, the toxicology subset of FSTA, *Pollution Abstracts, Toxicology Abstracts, Ecology Abstracts, Health and Safety Science Abstracts* and portions of *Aquatic Sciences and Fisheries Abstracts.* There are nearly one million items; POLTOX II: EMBASE provides 10 years of cover extracted from EMBASE and has over 300 000 entries; POLTOX III: CAB 1983– has about 170 000 records and, based on the CABI records, is more agriculturally orientated than the other databases.

General

Biological and Agricultural Index 1964– covers 225 journals, including 25 annual review serials, with a heavy US and Canadian bias. It appears monthly with quarterly and annual cumulations and is available as the CD-ROM BIOLOGICAL AND AGRICULTURAL INDEX 1983– , with 700 000 records, as a tape and online. A similar serial from the same stable is *General Science Index* 1978– , with a majority of articles from biological and medical journals. There are no authors' addresses.

Biology Digest 1974– is an abstract serial for high-school and college students, with a feature article in each issue. Papers are chosen from nearly 200 journals, predominantly American: there are both printed and CD-ROM versions. There are nine issues a year and an annual index of the about 3000 abstracts.

For SIGLE see p. 32. See also p. 104 for Indian databases. For SEC, RDIR and WORLD RESEARCH DATABASE see p. 35. For biotechnology and patent databases etc. see p. 133.

References

Wyatt H.V. When does information become knowledge? *Nature* 1972; 235: 86–89.

Wyatt H.V. *AIDS Information Supply and Demand*. British Library Research Paper 55. 1988; pp. viii and 67.

CHAPTER TWELVE

Other databases and CD-ROMs

H.V. WYATT

Since the third edition of this book there has been a revolution in the way that information is made available to individual scientists. They may now sit in their offices or laboratories and search databases through CD-ROMs in the library, or on diskettes on their own computers, storing complete references or abstracts on their own disks. This has many advantages, but also some disadvantages. CD-ROM versions of the secondary serials are considered with their databases in Chapter 11, p. 75. Some molecular databases are discussed at the end of Chapter 16.

One great advantage of CD-ROMs is that there are several systems for storing references directly from the databases. The systems can also be used to make one's own store and amalgamate it with those from databases. Some of these systems will also generate references, in either the CBE style for American life science journals or in Vancouver style for medical journals, and integrate them into the manuscript, avoiding the many errors in references which editors so bemoan, although errors in the system will be perpetuated as no-one will check the references. Seven of these systems have been reviewed by Jones (1993) (see also p. 51).

Databases and CD-ROMs

There has been an explosion in the number of databases in the life sciences. The number of large databases will increase and more will be available on CD-ROM and online. Meanwhile, individuals and institutions with computers will continue to produce smaller databases, first for their own use, then for colleagues, and finally for wider use. Libraries and museums have catalogued their stocks, individuals have produced bibliographies. There will be much duplication, but groups linked by e-mail may be able to use databases produced all over the world. It is therefore less useful to produce a list which will be out of date before printing.

Linking databases so that the information from one will be available immediately on another is the next step, although systems produced in isolation may not be compatible. The NSF Center for Microbial Ecology, (CME) at Michigan State University at E. Lansing is linking the Ribosomal Database, the Fatty Acid Methyl Ester Profile database and the Phenotypic Database in *Bergey's Manual* (see p. 178) and will add more databases in the future.

It is now so easy to process information and to make it available for searching and to publish, that many laboratories now have considerable amounts of accessible data. The Marine Biological Association laboratory at Plymouth might be typical, with the following projects: a British marine fishes database with details of more than 330 species; *Directory of Marine and Freshwater Institutions of Science and Research Engineering in the United Kingdom and the Republic of Ireland* 1992. (There is also a *Directory of European Aquatic Sciences Libraries and Information* Centres, edited by D.S. Moulder and N. Momzikoff (Paris, Institut Océanographique, 1991); *The Plymouth Marine Fauna*; a Marine Biodiversity network on the Internet; a library database including items of marine pollution worldwide; a database of archival material about the MBA, the laboratory etc.; a new edition of the library's serial holdings, 4000 titles, including report series of expeditions.

The library produces *Marine Pollution Research Titles* monthly and an annual bibliography, *Estuaries and Coastal Waters of the British Isles.*

The library is the UK national coordinating input centre for the *Aquatic Sciences and Fisheries Abstracts (ASFA).*

A series of directories by Bowker-Saur is now available: *World Databases in Medicine* (2 vols, 1993.); *World Databases in Agriculture* (1995); and, *World Databases in Life Sciences* (1996). All kinds of electronic database are to be included: online, CD-ROM, tape and diskette, e-mail etc., and there is likely to be considerable overlap between the different volumes.

The CD-ROM Directory Annual (TFPL Publishing) gave full details of 6000 titles in 1994; the disc version appears every six months and may be searched through Italian, French, Spanish and German interfaces. There is a directory of producers of CD-ROMs. SilverPlatter publishes an annual directory of its CD-ROMs which includes the GALE DIRECTORY OF DATABASES listing over 5300 databases, including 3500 on CD-ROM. For databases available in India, including Indian ones, see p. 104.

Databases on CD-ROM and disk

As it is relatively cheap to produce an update, a CD-ROM is now a favorite medium for large databases. The difficulty is that libraries are not able to buy as many as might be indicated by a listing of those available. CD-ROMs may also be stolen, a problem not foreseen by many. CD-ROMs can now be bought by a library and downloaded into a computer, where access from many licensed remote sites will be possible. This system enables very fast simultaneous searching of the databases and circumvents the problem of handling many disks at once by a machine. The problem remains as to how such services will be costed, and who will pay.

Some idea of the usefulness of a book may be obtained from a review, but few CD-ROMs are reviewed. *Nature* reviewed *Fishbase: a Biological Database on Fish*, a CD-ROM with a user's manual (*Nature* 1995; 376: 735). This has been compiled by the International Center for Living Aquatic Resources Management, Manila, and has details of the biology, distribution and taxonomy of 12 500 species of fin-fish, including all the (currently) commercially important species. The reviewers point out that this inexpensive – less than $100 – resource, although requiring a modern computer, is the sort of compilation from a widely spread literature, both published and grey, that will be so useful for developing countries which lack libraries and other information resources. We need to consider how such excellent new resources are reviewed and made known.

British Pharmacopoeia and *British Pharmacopoeia (Veterinary)* are contained on a single disc with addenda. Vidal EuroPharm is a collection of six databases from Switzerland, France, Italy, Germany, the USA and the UK *Martindale, the Extra Pharmacopoeia* and *Micromedex*. The databases are available separately or complete. ECDIN is a CD-ROM of about 7000 chemical substances which are produced in environmentally relevant quantities in the European Communities. It is drawn from the ECDIN database of more than 120 000 chemical compounds which can be searched online. It is a factual database which includes numerical values, and at least one expert checks each entry.

R & D Focus provides drug updates and abstracts for pharmaceuticals worldwide. *Release of drug data report* CD-ROM has information on 32 000 compounds.

The *CHCD Dictionary of Natural Products on CD-ROM* is a selection of 80 000 natural products and related compounds from the Chapman & Hall Chemical-Database (CHCD): this includes dictionaries of organic compounds, alkaloids, antibiotics, steroids and terpenoids, with material not published elsewhere. As with many other databases, search may be made by CAS Registry number, natural source and biological activity. Search by substructure is also possible. So far as I know, there is as yet no link between these chemical databases and the biological and genetic sequence databases.

AntiBase is a disc database of 17 000 natural compounds from microorganisms and higher fungi, including yeasts and lichens. Annual updates will include new information and compounds (see p. 40).

Reference

Jones R.G. Personal computer software for handling references from CD-ROM and mainframe sources for scientific and medical reports. *BMJ* 1993; 307; 180–184.

CHAPTER THIRTEEN

Libraries, museums and other resources

H.V. WYATT

Most countries now have a national system whereby libraries may borrow books and journals from each other; many library catalogues may also be accessed online. Eight UK university libraries and Trinity College Dublin form a Consortium of University Research Libraries (CURL), which is cooperating with the North American Research Libraries Group (RLG). The groups are expanding and hope to produce a database of catalogues commercially available online. When a national network cannot find requested books or journals, loans or photocopies may be requested from other countries. French university library catalogues are now online and used for searching and loans. The interlending OSI Network (ION) links libraries in the UK, France and The Netherlands. Libraries are still cancelling subscriptions to journals, thereby making interlibrary loans even more important. Recent cuts at Yale University prompted a call for authors to send reprints, so that a database of the *Escherichia coli* linkage map could be maintained.

A Middle Eastern library at King Fahd University of Petroleum made 1340 loan requests in a year, mostly in engineering. Nearly 40% were made to BLDSC and 15% each to Universiatsbibliothek und TIB and to Bibliothek TU Delft (Siddiqui, 1990). Training in another country, friendships and visits abroad will affect the use of different sources.

English-language journals are all-powerful and form the majority of journals taken by libraries, even in French-speaking countries: for example, in Tunisia a research library takes 150 journals and serials of which only 20 are in French. The INIST collection (see below) is 40% English language and 16% French.

In medicine there are enough doctors and lucrative advertising to support journals in many different languages, e.g. the highly respected *Nederlands Tijdschrift voor Geneeskunde.* Many papers are published in English in an international journal and then in a national journal in the respective language for the benefit of the local doctors. It is annoying to obtain such a paper after much trouble, only to find that it contains the same information as the English language paper. This is probably a minor problem in the biological sciences.

Many countries have national libraries for medicine and agriculture, but the life sciences are usually part of the general science collections. There may be collections associated with specific societies, such as the Zoological Society in London, but the libraries may be restricted to members and the collections may be lacking in subjects such as biotechnology and biochemistry. However, the larger and older universities may have excellent collections. For his book *The Emergence of Bacterial Genetics* (1990) T.D. Brock used numerous archive collections and personal correspondence with scientists. He wrote: 'all the library work was carried out in the libraries of the University of Wisconsin, Madison. I can think of no better place to work on a book such as this, because these libraries are absolutely superb in their holdings of 19th and 20th century material in bacteriology and genetics'.

The major powers hold libraries in the capitals of most countries, and in some large cities. There are British Council libraries in most countries: most will have the bibliographic database BOOKFIND on CD-ROM, with details of over 500 000 books in print and others out of print. In Kenya, for example, there are libraries in Nairobi, Kisumu and Mombasa: BOOKFIND is available in all, but the MEDLINE and CABI CD-ROMs are available only in Nairobi, whence the results of searches can be sent to other libraries. The libraries also accept requests for loans or photocopies from BLDSC (see p. 98), but unfortunately the charges are high and so people may obtain copies from other libraries, e.g. from France. For medical articles the BMA Library in London is cheaper than BLDSC.

The Educational Low-Priced Books Scheme (ELBS) made unabridged British textbooks available to students in 54 countries at one-fifth to one-third the price of the cheapest standard editions. Two catalogues, *Medical and Nursing Books* and *Science and Agriculture Books*, were issued every two years. In 1995 the UK government announced plans to end this scheme, but the outcry may force a modified scheme to be retained. Inclusion of a book in these catalogues implied a very useful text. The USA also supplies some books at subsidised prices.

The quarterly select list of latest additions to British Council libraries is widely circulated. *British Medicine* is a valuable tool. The British Council supplies two new British journal titles free for two years to research institutes. Any library may become a corporate member of the British Council library for a small fee, and may borrow 100 books for three months. Some libraries circulate abstracts from British journals received, most by air mail, to interested scientists. Photocopies of articles may be bought from the library or from BLDSC. Research institutes and colleges etc. receive gifts of substantial numbers of new British books. Unfortunately, many institutions have an impressive display of journals received as gifts, but have been unable to buy any since. There are several groups of British academics who pass on journals to poorer countries.

The International Network for the Availability of Scientific Publications (INASP), under the aegis of International Council of Scientific Unions (ICSU), supports and encourages the exchange and donation of books, journals etc., offering a clearing-house and advisory service. There is a newsletter and an annual *Directory*, with very useful listings of agricultural, biological, forestry, medical, marine sciences and veterinary organizations all over the world, together with (donor) organizations specializing in information development. However, the Scientific Apparatus Recycling Scheme (SARS) of FEBS is listed under science, not biology.

Guides to libraries

Most guides to libraries are disappointing and of little of use to life scientists, apart from addresses, e.g. *World Guide to Libraries* (11th edn., Bowker-Saur, 1993). *World Guide to Special Libraries*, edited by B. Bartz, R. Lochar and H. Opitz (2nd edn., 2 vols, Bowker-Saur, 1990) is useful. The following British Library guides are useful, if only for addresses, telephone and fax numbers, but have limited information for life scientists: *Guide to Libraries and Information Units in Government Departments and Other Organisations*, edited by P. Dale (32nd edn., 1996) is a guide to about 600 libraries in the UK only. It includes agriculture, but not academic libraries. There have been so many changes in government and other organizations that it is necessary to consult the current edition. *A Directory of Information Sources in Biology in the UK* by H.V. Wyatt (BL R&DD Report No. 5606, 1981) gave details of 77 organizations, which have since suffered many amalgams, closures and changes. However, the directory listed the start of the main collections, as opposed to the more usual earliest volumes, with sometimes several hundred years' difference, and current journals, cultures, slides, paintings etc. as well as special search facilities. This directory is therefore still a useful reference when used with a current directory with less detailed information. *Guide to Libraries in Western Europe. National, International and Government Libraries*, edited by P. Dale (2nd edn., 1994) has details of some fishery and agricultural libraries. *Guide to Libraries in Central and Eastern Europe*, by M. Hughes (1992): the Central Agricultural Library, Warsaw, and the National and University Library in Ljubljana both publish national bibliographies. *Guide to Libraries in Key UK Companies*, edited by P. Dale (1993) lists some few food, water and pharmaceutical companies, but there is often very limited access to company libraries as even knowledge of the current journals may be a clue to new developments.

More useful is the *Directory of Libraries and Book Agents in the Former Soviet Union and Eastern Europe*, edited by R. Hogg and K. Ladizesky (BLDSC, 1994). In addition to the addresses in the east, there are very useful addresses of agencies in the UK, USA, France and Germany. There is also a list of Cyrillic titles and a list of the cover-to-cover translations of Cyrillic titles. There is a list of Slavonic titles and an organization index (see also p. 109).

The *Directory of Australian Academic and Research Libraries*, edited by A. Bundy and J. Bundy (5th edn., Adelaide, Auslib Press, 1992) lists every conceivable library but contains little information for the scientist. The importance of the collections can only be judged by the number of CD-ROMs listed.

The *Directory of Special Libraries and Information Centres*, edited by J.M. Zakalik (2 vols, 17th edn., Gale) has 21 380 entries, mainly for the USA and Canada. It is very good for looking up a particular institute or facility, e.g. germplasm or *Drosophila* collection, but is useless for finding a subject such as biotechnology or microbiology as there are so many entries and little indication of their size or importance.

Life Sciences Organizations and Agencies Directory (LSOAD), edited by B.T. Darnay and M.L. Young (Gale, 1988) has 8000 entries, including 860 libraries and information centres in the USA and 184 in Canada. Details are given of 70 botanic gardens and arboreta in the USA. The other entries are also mainly from the USA, with some from Canada. Associations are US and international, as are computer information services. There are 290 US and 31 Canadian consulting firms. The *Research Centers* sections gives details of 1500 universities and 167 research services in the private sector in the USA. Another 1633 research centres are listed from 100 countries, Algeria to Zimbabwe. Eighty organizations concerned with standards are given. Details of over 300 US state and federal government agencies and research centres are listed. This is a very good directory with a good index. A more recent *Research Centers Directory 1996*, edited by A.L. Gerring (2 vols, 20th edn., Gale) published in 1995, covers only the USA and Canada, and only university and independent non-profit institutions. There are more than 900 entries for agriculture, food and veterinary science, 1400 in biology and environmental sciences, and 2250 for medicine and health sciences. There are detailed entries on the activities of each department etc. Two companion volumes are *Government Research Directory* and *Research Sciences Directory* for profit making firms and individuals.

Less satisfactory is *Biosciences. Information Sources and Services*, by Y. Alston and J. Coombs (Stockton Press, 1992). The international, national and government organizations section has very limited accounts of each and is very selective. There is a long section listing biological journals but the selection seems arbitrary, e.g. *Journal of Experimental Medicine*, one of the highest-ranking immunological journals, is not included.

In 1993 a small low-orbiting satellite provided the first low-cost electronic link between health institutions in Africa and cooperating sister libraries in Europe and North America. HealthSat II links 18 ground stations in Africa, from Botswana to Ethiopia, with experts and libraries for a two-way exchange of queries and information. The satellite passes every spot on earth at least four times a day. SatelLife is a non-profit making charity started by the International Physician for the Prevention of Nuclear War with the money from the 1985 Nobel Peace Prize. Its first use was to transmit to Nairobi a paper on the treatment of measles with vitamin A, research made in Africa but, because of lack of library funds, still not known in Africa six months after its publication in *New England Journal of Medicine*. So far the link is only for health use, but it represents an exciting possibility for similar satellites and cooperation devoted to agriculture and other life sciences. (see p. xv of the Introduction).

Some important European libraries

The Food and Agriculture Organization (FAO) Library, Rome, has one million volumes and 7000 current serials. It adds about 2500 monographs a year. FAOBIB is a database of FAO technical documents 1945– and library monograph holdings 1975–. It is good for fisheries, nutrition, statistics and forestry, and is the international centre for ARGRINET, with loan services, document delivery and SDI services. AGRINET is a worldwide network of large agricultural libraries. FAO documents may be puchased in full-text microfiche. The Current Agricultural Research Information System (CARIS) lists 27 000 active projects in English, French or Spanish, with a multilingual thesaurus called AGROVOC (2nd edn., 1992) in German, Italian, Portuguese and Arabic versions. There is a CARIS CD-ROM 1995, and other access via print, online, tapes and diskettes. FAO also produces AGRIS and associated products (p. 86).

There is an Information Centre for Land, Soil, Water and the Environment at Wageningen NL-6700 AA Netherlands. The International Council for the Exploration of the Sea in Copenhagen Denmark, has collections on fisheries, oceanography, marine pollution and the environment.

The National Medical Library for the Federal Republic of Germany, Cologne D-50924, has an extensive biomedical collection and makes interlibrary loans.

The Institut de l'Information Scientifique et Technique (INIST), at Vandoevre-les-Nancy, is the French equivalent of BLDSC (see below). It holds 27 000 journal titles, of which 19 000 are current, 40 000 reports, 50 000 conference proceedings and 64 000 French doctoral theses in science since 1980. Biology forms about 20% of the collection and medicine about 15%. PASCAL is the major secondary serial, with about 50% devoted to the life sciences and medicine. From 1973 there are 10 million records, increasing by 600 000 a year (see p. 79). PASCAL is online and INIST will translate items. There is a special Japan Unit. Public health and the history of science are on the smaller FRANCIS database. INIST is linked to other French libraries for items not in the collection. The Institut Pasteur in Paris has a catalogue which is good for French and foreign journals and serials. There is also a good archive collection.

The British Library (BL) is the UK's national library. For scientists there is the BL Document Supply Centre (BLDSC) at Boston Spa, Yorkshire, and the Science Reference and Information Service (SRIS), which includes patent services, in central London. BLDSC provides photocopies and lends books and journals where these are required, and searches for documents held elsewhere. In 1994 some three million British and nearly one million overseas requests were received. Requests for loans etc. from overseas may be made through a registered library. There is a small reference collection for visitors and a very limited service for requested material; many of the secondary serials are not housed in the reading room and reference material may be on loan.

SRIS has its life sciences collections at Kean Street, London WC2B 4AT, and has a large reading room with a large reference section and many current journals on open access. Books and serials may be requested. There are extensive facilities for online searching etc. Help with foreign languages is available on certain days. Patents are housed in a different building close by. Both SRIS and the patents should move to the new British Library complex next to St Pancras Station in the near future. Table 13.1 shows the resources at these two libraries.

The Science Museum Library in London has 600 000 volumes from the 15th century onwards. It now concentrates on providing a first-class resource for the history and public understanding of science. It has very rich pictorial and archive collections, with company records, personal papers etc. There is a full set of British patents. The Natural History Museum has the largest and most comprehensive book collections on natural history and the life and earth sciences in the world, with over one million volumes, including 800 000 books. Special features are major works beginning in the 15th century, with unique and rare copies, some signed and annotated, and supporting manuscript material. There are half a million original drawings, paintings and prints, including over 100 000 original works of art, the third largest collection of original watercolour drawings in Britain. The General and Zoology, Botany and Entomological Libraries are at the main museum site in London SW7, while the Ornithology and Rothschild Libraries are at Tring in Hertfordshire The museum has extensive collections of animals and plants, maps etc. The Linnaean Society Library in the centre of London has an excellent library. About 900 journals are taken. These include many newsletters etc. from all over the world which are unlikely to be found in other libraries. There is a large historical section with manuscripts and special collections. Kew is a major resource for botany (see p. 239). The Wellcome Institute for the History of Medicine, London, has about half a million items, many of which cover microbiology, biochemistry, malaria, nutrition, tropical medicine etc. Societies such as the Biochemical Society of the UK and the American Society for Microbiology, have historical collections and taped interviews.

Libraries in the Americas

The Library at Woods Hole, Massachusetts combines the resources of the Marine Biological Laboratory and the Woods Hole Oceanographic Institution, with nearly a quarter of a million volumes and 5000 journal titles in biology. Journals may only be used in the

Table 13.1. The resources of British Library at the Science Reference and Information Services in London and at the BL Document Supply Centre in Yorkshire. Stock at the beginning of 1995.

SRIS			BLDSC	
Stock	Annual		Stock	Annual
33 000 000		Patents		
3 000 000		Monographs, books	3 000 000	42 000
3 400 000		Reports on microfilm	3 850 000	110 000
	67 000	Journals	247 000	47 000
		PhD theses USA	450 000	6 000
		UK	110 000	6 000
		Conference proceedings	335 000	16 000
		Translations	560 000	9 000
		Other reports	500 000	25 000
		Cyrillic script books	220 000	1 000
		Cyrillic current titles		4 000
		Microfiche other than reports	300 000	
		Roll microfilm	1700 miles	

library so that, like the SRIS in London, it is ideal for concentrated literature searches and use. The library has reciprocal lending agreements with 80 libraries and the Boston Library Consortium for photocopies and books. The resources of the library, including bibliographies and databases, may be accessed through the Internet (Woods Hole Scientific Community gopher).

The National Library of Medicine (NLM) at Bethesda, Maryland, is the major medical library in the world and is best known for its secondary serial *Index Medicus.* The library has over five million items, of which over two million are books, and is at the centre of the (US) National Network of Libraries of Medicine (NN/LM) of 4000 libraries, fulfilling more than 2.5 million interlibrary loans each year. NLM is primarily for the medical community, but this is interpreted very widely: in subjects such as microbiology and biochemistry and molecular biology the collection and incorporation into the *Index Medicus* database is catholic. It is therefore a major resource for life scientists, and because of the skills of its operators should always be carefully considered for use as the secondary serial of choice. There is an excellent History of Medicine section. Like the Wellcome collection in London, NLM collects a wide range of material from the life sciences related to medicine, including personal papers etc.

The National Agricultural Library (NAL) at Beltsville, Maryland, is supplemented by 12 USDA field libraries across the country. About 25 000 serials are received and a selection of 600 are contained in the Periodical Reading Room. There is an extensive collection of documents of the UN, FAO, USDA and other US government documents on microform, as well as the reports from agricultural experimental stations. The library, the largest agriculture library in the world, has over two million items, of which 60% are from abroad. A branch library in Washington DC has more than 250 000 items. The Food and Nutrition Information Center has its own collection, which includes extensive teaching material. AGRICOLA is the bibliographic database of NAL (see p. 86) from which is printed *Bibliography of Agriculture*. NAL bibliographies include *Transgenic Fish Research* and *The Value of Wetlands*. Also based at the NAL are the Plant Genome Data and Information Center, Aquaculture Information Center and the Biotechnology Information Center. The pig genome database *USPIGBASE* can be accessed via the Internet, as can the plant genome database, which includes tomato, maize, rice, soybean, small grains such as wheat, barley and oats, and *Arabidopsis*.

The National Museum of Natural History, a branch of the Smithsonian Institution in Washington DC, has major collections of books and receives nearly 4000 journals. There are nine smaller libraries, like other similar institutions, the Museum has extensive collections. The insect collection of 17 million specimens is the second largest in the world, and there are also 12 million invertebrate and four million vertebrate specimens. There is a branch library at the Tropical Research Institute in the Republic of Panama, with special emphasis on the fauna and flora of tropical areas, especially Panama and adjacent areas. Other important libraries are the American Museum of Natural History at Central Park in New York, and the Academy of Natural Sciences of Philadelphia, Pennsylvania. Many of the large public and private universities have magnificent libraries (see above).

Latin American Research Libraries in Natural History – a Survey, by T.J. Davis (American Ornithologists Union, 1989) is recommended.

India

India has one quarter of the world's scientists but its contribution to the literature is small, with much published in Indian journals of below-average quality. Indian science is not well represented in the secondary serials or in library collections outside India. The National Science Library is at INSDOC in New Delhi, and is well organized, with one of the largest collections of Russian books and journals in Asia, e.g. 30 000 books and 600 journals. There is access to all major databases. The *Medicinal and Aromatic Plants Abstracts (MAPA)* 1979– which uses 600 journals in 55 languages, is also available online 1990–: *MAPA* is particularly good for information from other Asian countries. INSDOC also provides, a 48-hour express service for contents, abstracts or photocopies based on 1000 core journals divided into high-value and high frequency journals and low-value and low-frequency journals, as well as a service based on *Chemical Abstracts*.

Indian Science Abstracts 1964– covers more than 25 000 articles in Indian journals each year; the database may be searched. *NCI Current Contents* 1992– appears monthly and gives contents pages for all Indian science and technology journals, including agriculture and medical sciences. There are author and title word indexes, but no addresses. There is an online version.

Indian libraries are now pooling their subscriptions to foreign journals following large cuts in library budgets: the science library of New Delhi University bought 795 foreign journals in 1992, but only 275 in 1994. Libraries cooperate with a photocopy service and local delivery in the large centres. Some institutes have a unified subscription policy and rotate current journals. INSDOC has produced the following reference works: *Directory of Indian Scientific Periodicals* (4th edn. 1992) with additional data about research and industrial institutions; *Directory of Scientific Research Institutions in India* (6 vols, 1989); and *National Union Catalogue of Scientific Serials in India* (4 vols, 1988) gives details of holdings of 840 libraries: the data are available online. Many holdings may not be as complete as might be expected from this catalogue. The *National Index of Translations* is a database of about 20 000 translations since 1981, and is being extended to include all those held in India. The *Indian Science Citation Index (ISCI)* 1990– is a database.

INSDOC has a database of 40 000 Indian patents 1972– and the National Chemical Laboratory at Pune (Poona) has a bibliographic database on Indian chemical patents, producing *Indian Patent Bulletin* 1975–1987, with cards which may be searched, and *Indian Chemical Patent Abstracts* monthly 1987–. This latter adds about 1000 patents each year and includes agricultural and medical chemicals, bio-engineering and food and agriculture.

The *SCI* database includes 11 Indian biological journals: *Acta Botanica Indica*, *Comparative Physiology and Ecology*, *Enomon*, *Indian Journals of Animal Science, Biochemistry and Biophysics, Experimental Biology, Genetics and Plant Breeding* (four separate journals), *Journal of Biosciences*, which has nearly 20% of its articles from abroad, *Journal of the Indian Institute of Science C:*

Biological Sciences, *Proceedings of the Indian Academy of Sciences* in both *Animal Sciences* and *Plant Sciences* series.

There are a number of good natural history journals. Theses range from the excellent to very poor: abstracts of theses are published by some universities. *Current Science* appears every two weeks and is a major international journal with many very interesting special issues, e.g. Pollination biology in the tropics, and excellent colour illustrations. There are a few book reviews, but more important are the listings of books received, usually books published in India.

Nearly 20 Indian medical journals are published as full text and image on the CD-ROM *ExtraMED* (see p. 108). Many Indian doctors work in the Gulf area and publish in the local journals, some of which also feature on the CD-ROM.

The Distributed Information Centre (DIC) in Bioinformatics at the University of Poona is one of nine biotechnology DICs with direct access to all major databases and databanks worldwide, with the ability to analyse information and model. Eighteen centres will be linked, with access to journals, papers, reports, patents etc. The Pune centre has produced a databank, Animal Virus Information Systems (AVIS), of more than 600 viruses, and a Protein Secondary Structure (PSS) Databank. The DIC on enzyme engineering, immobilized biocatalysts, microbial fermentation and bioprocess engineering is at the Institute of Microbial Technology (IMT), Chandigarh, which also has the national Microbial Type Culture Collection, with 3000 cultures and gene and plasmid banks. A catalogue was published in 1992. The National Collection of Industrial Microorganisms is kept at the National Chemical Laboratory, Pune, and collections of fungi, blue-green algae, plant tissue cultures and germplasms are held at the Indian Agriculture Research Institute, New Delhi. Several hundred isolates from Indian oilfields are kept in the Zoology Department at the University of Poona. Food organisms are kept at the Central Food Technological Research Institute at Mysore.

Biotechnology has received considerable funding and has attracted many Indians to return from the west. Many of the modern centres in India now have e-mail facilities. The International Centre for Genetic Engineering and Biotechnology was opened in New Delhi in 1995 (the other centre is in Trieste, Italy). The National Bureau of Plant Genetic Resources (NBPGR), New Delhi, has a seed bank and is linked to the International BPGR of the FAO. Based at the Indian Agricultural Research Institute, it will have facilities for 800 000 accessions of seeds etc. – comparable with Fort Collins, Colorado. A national gene bank of all economic plants is to be established in the Tropical Botanic

Table 13.2. Publications on poliomyelitis in India (from Wyatt, 1992)

Journals	1970–1979	1980–1989	Total 1970–1989	Entries in *Tropical Diseases Bulletin*
Indian	94	127	221	8*
International	11	43	54	14

* referred to a symposium of nine abstracts.
The 54 papers in international journals appeared in 22 journals. Only 37 items dealt with rehabilitation or surgery. Very few of the papers in Indian journals appeared in *Index Medicus*.

Gardens in Trivandrum. There is a tree germplasm bank at Hissar in Haryana, with a planned total of 4000 woody species. The International Institute for Arid and Semi-Arid Technology, ICRISAT, at Patancheru near Hydrabad, has germplasm collections for sorghum, pearl millet, pigeon pea, chick pea and groundnut.

The National Institute of Nutrition at Hyderabad hosts the Laboratory Animals Information Service Centre, which has a library and produces a newsletter. *Current Research in Plant Science (CRIPS)* 1990– is produced by the National Botanical Research Institute, Lucknow. *Paryavaran Abstracts* 1984– covers current Indian literature in the environmental sciences.

The British Council has libraries in several large cities. The library in Calcutta, for example, has 14 journals of general and environmental interest and 14 medical journals of interest to biologists. *Index Medicus* is available on CD-ROM.

The Handbook of Libraries, Archives and Information Centres in India, edited by B.M. Gupta and S.S. Nathan, *Asia-Pacific Cooperative Information Systems, Networks and Programmes* (vol 4, Information Industry Publications, New Delhi, 1987) and *Science and Technology Information Systems and Centres* (vol 7, Adilya Prakashan, Delhi, 1990) may have information for biologists. UNESCO has an Asian network for biological sciences.

Sixty four databases in India have been identified (Chavan and Chandramohan, 1995). Forty-seven are bibliographic or directory in nature, of which 33 cover Indian data. Of 17 numeric etc. databases, 12 cover Indian-only sources. Most are available only in print or floppy disk format. UNESCO has published *Databases in SE Asia/Pacific* 1990, but its usefulness is very limited as it appears to have been compiled in 1985.

Indian promotion committees show poor regard for their own journals and, publication in international journals is therefore encouraged. Authors are thus encouraged to compete with western research, and to neglect purely Indian themes which would be unlikely to be accepted for western journals. International journals and the major secondary serials are far too expensive for most libraries, and so Indian research published in international journals becomes inaccessible to other Indian scientists. An example of this bias is shown by poliomyelitis, which in the period 1970–1989 probably resulted in about four million cases of paralysis (Table 13.2). In spite of this enormous morbidity, there has been little interest and fewer than 300 editorials, notes, letters, annotations and papers in this time.

Museums

Museums are an important resource for animal and plant specimens. *Biological Collections UK* (Museums Association, 1987) is not, as its title might imply, a guide to museums in the UK, but is a report of a survey of nearly 300 institutions out of 604 with collections. Almost all had zoological and two-thirds had botanical specimens. There is a list of institutions, but this is clearly not inclusive. The report concluded that many collections were not viable, with small libraries and no curator, unsafe, rotting collections and no visitors. There is a list of a few institutions which might take over old collections. There was a general contraction of museum services in the 1980s, which is probably continuing. Some municipal museums in Europe and North America have extensive and important collections: Liverpool Museum, for example, has the most extensive collection of timber specimens in the UK, as a result of its prominence as an importing city. Details of collections held in museums may be found in some research newsletters (see p. 27). *Heteropterists Newsletter* published a supplement with details of 115 collections in 35 countries. *Tymbal* noted that North Carolina State University has 250 000 specimens of Homoptera and important holdings of literature. The use of the Internet may make databases of collections easier to establish.

A national database at Manchester University has details of 20 000 life science collections in the UK. It is thought that more than half the type specimens are held in local rather than national collections. Of the more than eight million mollusc specimens in the UK, nearly half of those from abroad are held in 32 local museums. The major collection in the UK is the Natural History Museum with, for example, 22 million insects. The American Museum of Natural History in New York is the world's largest resource for birds, spiders, fossil mammals and whale skeletons, and has other enormous collections.

Natural History Manuscript Resources in the British Isles, by

Q.D.R. Bridson, V.C. Phillips and A.P. Harvey (Mansell, 1980) and *A Catalogue of Natural Science Collections in North-East England with Biographical Notes on the Collectors*, edited by P. Davis and C. Brewer (North of England Museum Services, 1986) are useful.

References

Chavan V., Chandramohan D. Databases in Indian biology: the state of the art and prospects. *Current Science* 1995; 68: 273–279.
Siddiqui M.O. Interlending in a university library: The KFUPH experience. International Library Review 1990; 22: 329–338.
Wyatt H.V. Indian medical journals. *Lancet* 1992; 340: 613.

CHAPTER FOURTEEN

Foreign-language literature and translations

H.V. WYATT

English is the major language for the life sciences and the majority of important papers are published in North American and European journals. Even in clinical medicine, where many countries publish society journals in their own languages, the impact of such journals is small. Table 14.1 reflects the number of papers from the countries indicated that are published in the national journals, not the total number of papers. Most will have been published in the national language, but some may come from other countries. The Université des Réseaux d'Expression Française (UREF) is an association of 200 French-speaking universities which publishes journals, CD-ROMs and books, e.g. *Manuel de Techniques Virologiques.* There are regional offices in Canada, Madagascar, Haiti and Senegal.

Current Contents is a very useful way of looking at journal contents, but it is only a selection of the most heavily cited journals and excludes many journals which are important outside the mainstream of western science. A glance at the subject lists in *Journal Citation Reports* shows that many of the lower-ranking journals are non-English and should not be neglected.

The Library of Congress has six field offices, in New Delhi, Cairo, Karachi, Jakarta, Nairobi and Rio de Janeiro, to buy and ship books and other material from 34 countries to 87 libraries in the USA and to libraries in nine other nations. From 1962 to 1991 New Delhi has shipped nearly 17 million items, including 137 medical journals in English and 16 in other Indian languages from India alone. Unfortunately, there is no separate breakdown for either life or agricultural sciences. For particular items it might be

Table 14.1. Ranking of mainland European journals (national), by three or more citations in *Current AIDS Literature* July 1987 – September 1988 (from Wyatt, 1988).

Country	Journal ranking	Number of citations
West Germany	6, 18, 112, 137, 137	172
France	48, 65, 74, 82, 82, 100, 137	47
The Netherlands	17	44
Italy	25, 74, 137	37
Switzerland	45	17
Austria	137	3
Belgium	137	3

worth contacting the Library of Congress for help in locating likely recipients of materials.

WHO has promoted some *Regional Index Medicus* projects (see p. 56), but a new venture, *ExtraMED* 1994– (Box 1359, London W5 3ZP, UK) is a monthly CD-ROM with full text and images of 200 biomedical journals. Most of the journals are from China, Africa, the Indian subcontinent and the Pacific; unfortunately, the list of journal titles does not include the country of origin unless it is in the title itself. A few journals will interest life scientists, e.g. *Bangladesh Journal of Botany, Bulletin of the Nutrition Institute, Chinese Journal of Medical Genetics, Chinese Journal of Microbiology and Immunology, Egyptian Journal of Food Science, Indian Journal of Malariology* and *Pakistan Journal of Biochemistry.* The cost, at £10 a journal, is not high, but probably only specialist libraries will be able to afford this service. It remains to be seen whether FAO or other sponsors will initiate similar projects of more interest to life scientists.

Translations

Important papers are often published in English as well as in the national language. The French have finally yielded, and the famous *Annales de l'Institut Pasteur* are now published in English under new titles. MEDLINE contains 111 journals of French origin. Nevertheless, important papers are published in other languages and are not easily accessible. The International Translations Centre (ITC) in Delft, The Netherlands, is a clearing-house for translations of scientific and technical documents into western languages. Over 200 organizations send details of translations to ITC, which publishes *World Translations Index* (WTI) of about 30 000 translations each year and their locations, with an annual cumulation: there are five- and nine-year cumulations for 1967–1985. The entire WTI database of 400 000 translations from 1977 may be searched online or on magnetic tape. Two-thirds of the translations are into English, and of these a half are from Russian and a third from Japanese and German. However, less than a quarter of the translations are in the life sciences and agriculture. Over three-quarters are from journals, but 15% are of patents, monographs and conference proceedings.

In the USA the John Crerar Library of the University of Chicago houses the National Translation Center. The *Consolidated Index of Translations into English II 1967–1984* is available in print and in microfiche. There are also the *Translations Register Index* 1967– and *Transdex Index* 1974–.

Journals in Translation (6th edn. 1996), listing 1300 titles, is published by ITC and BLDSC. The monthly *British Reports, Theses and Translations (BRTT)*, also from BLDSC and available on BLAISE, is also useful. There are quarterly cumulations of keywords and authors on microfiche. The 1995 issues contain theses from Cambridge from 1990 and 1991 as well as reports etc. from the Overseas Development Institute, the International Institute for the Environment and Development, the Institute of Terrestrial Ecology at Bangor, Wales, the Forestry Commission, English Nature and the Environmental Economics Centre, for example. There are monthly subsets: *Focus on British Biological and Medical Sciences Research* and *Focus on British Environmental Sciences Research.*

Russian Journals

Consultants Bureau, through Plenum, publish 12 Russian life science journals in translation by a team working closely in Moscow with Russian editors and authors. The translated journal appears concurrently with the original journal:

Applied Biochemistry and Microbiology;
Biology Bulletin;
Doklady Biochemistry;
Doklady Biological Sciences;
Doklady Biophysics;
Doklady Botanical Sciences;
Microbiology;
Russian Journal of Developmental Biology;
Russian Journal of Ecology;
Russian Journal of Genetics;

Russian Journal of Marine Biology; and
Russian Journal of Plant Physiology.

The Academy of Sciences of the USSR and Turpin Transactions publish jointly *Biomedical Science* 1990– in English: papers are refereed by both Russian and international experts. Turpin Transactions also publishes *Index of Current Medical Literature in the USSR* 1989–, which contains not only journal articles but also grey literature, patents, conferences and trade information. Both these publications cover a wide range of subjects, such as biochemistry, virology, immunology etc.

Soviet Medical Reviews: Sections D and E, respectively *Immunology Reviews* 1988– and *Virology Reviews* 1987–, are in English and available through Gordon and Breach and Harwood Academic Publishers. *Biological Membranes*, renamed *Membrane and Cell Biology* from 1996, is a translation of *Biologicheskie Membrany. Soviet Scientific Reviews: Section F. Physiology and General Biology Reviews* will cease in 1998.

Japanese journals

The JICST-E file 1985–, which covers science, technology and medicine in Japan, is a comprehensive bibliographic database with English citations and abstracts. It is based on about 4000 Japanese journals and serials, conference proceedings, technical reports and Japanese government and local government literature. It is updated every two weeks with about 10 000 citations, and contains over two million items. Also listed are research projects, institutions and staff. It is available online, with several hosts. Current-awareness searches may be made by and at the SRIS, London, UK (see p. 98), where linguists will help with any language difficulties. An interesting insight into Japanese was 'How to read Japanese medical journals' by P. Skrabanek (*Lancet* 1983; 2: 1482–1484.).

Chinese journals

Chinese Science Abstracts Part B 1982– (Pergamon) includes abstracts from about 20 life sciences journals but has no author addresses. Bibliographies and reviews of particular subjects, mainly parasitology, are published in *Tropical Diseases Bulletin*, e.g. a 50-page review with 300 references on *Clonorchis sinensis* in volume 91, 1994. *Biomedical and Environmental Sciences* is an international journal with special emphasis on scientific data and information from and published in China, and distributed in the west by Academic Press. The *Directory of Chinese Agricultural and Related Organizations*, edited by Zhang Qiaoqiao (CABI, 1994) has nearly 1000 entries and a checklist of relevant Chinese journals. There are about 700 Chinese medical journals, mostly in Chinese, although some have titles and abstracts in English. Few are included in *Index Medicus*, but about 60 Chinese medical journals are available as full text and image on *ExtraMed* (see above).

India has a collection of Russian journals and books at INSDOC and a collection of translations (see p. 101).

Reference

Wyatt H.V. *AIDS Information Supply and Demand*. British Library Research Paper 55. 1988; pp viii and 67.

CHAPTER FIFTEEN

Genetics

M.G. BURDON

The ways in which offspring resemble, or differ from, their parents has intrigued mankind from time immemorial. Fundamentally, the science of genetics attempts to provide a rational and systematic basis for what is otherwise little more than a collection of unrelated observations and anecdotes. At a more serious practical level, it has long been known that some diseases or conditions 'run in families', and the area of medical genetics has now become a central part of the subject. In reality the science of genetics underpins the whole of biology, in that the mechanisms that determine the processes of inheritance also underlie biological processes such as reproduction, development and evolution. Indeed, the various branches of genetics deal with the ways in which the principles of genetics are applied, and these divisions form the basis for the later sections of this chapter. However, it would be quite wrong to suggest that genetics is a collection of unrelated areas, particularly as the explosion in our understanding of molecular genetics has provided an elegant unifying basis for the whole subject. Developmental genetics, for example, has benefited recently from some particularly striking insights at the molecular level.

Classical genetics had a long and distinguished history before there were any real insights into the molecular mechanisms underlying inheritance. As a result of this basic ignorance of the mechanisms involved, geneticists had to deal with the organism they were using as a 'black box', whose contents were unknown and which could not be dismantled for examination. The only feasible approach to gaining information about the contents and organization of the box would be to arrange for a series of input signals to be fed into it, and then to measure the output signals and deduce as much as possible about its contents. Classical genetics works in a rather analogous way, where the input signals are the mutations that are fed into the experimental crosses and the output is the phenotypic changes, offspring ratios and so forth that can be observed. Initially, as was the case with Mendel's work, the input mutations were confined to naturally occurring variants, but subsequently, with the discovery of induced mutation by physical and chemical agents, the range of different mutations became virtually limitless. Most plant and animal breeding has been done in this way, crossing strains that appeared to have promising properties and looking for desirable progeny. This process does not in principle require a fundamental understanding of the mechanisms involved, and indeed much has been done without any such understanding. Although recent developments in genetics clearly hold great promise for the breeding of new strains of plants and animals, the plain fact is that this is a technology of the future, and almost all the achievements to date have been made by classical genetic means.

The transition from classical to modern genetics has taken place in two stages. The first, in the 1940s and 1950s, revealed the nature of the genetic machinery and led to a rapidly developing insight into how it functioned. The proof that DNA is the repository of genetic information, followed by the elucidation of its structure and the consequent insight into how such a molecule could replicate itself, meant that geneticists now had a good idea about what was going on inside the black box at a molecular level, even if their ability to manipulate it was for a long time strictly limited.

The second stage was, of course, the development of recombinant DNA (genetic engineering) techniques in the early 1970s. Suddenly our electronic engineer had been given a screwdriver to open the box, as well as the equipment needed to modify and rearrange the components inside. It is this combination of basic understanding of the molecular mechanisms and ability to carry out complex and far-reaching modification of the genetic apparatus that has made genetics such a powerful (and to some minds awesome) science. This manipulation of genetic material *in vitro* has now become so routine that many problems are now tackled directly by DNA cloning, amplification and sequencing techniques, thus standing the traditional experimental approaches to genetics on their heads. This has given rise to the concept of 'reverse genetics' to describe the process of working outwards from the gene to the phenotype, rather than the other way round as with the classical genetic approach.

Molecular genetics provides the unifying theme that underlies virtually all branches of genetics. However, when it comes to surveying the literature on genetics it is inevitable that both books and, in particular, journals will tend to reflect the time-honoured divisions of genetics, at least in their titles if not their contents. For example, although the two journals *Genes and Development* and *Molecular Microbiology* clearly deal with different areas of genetics, much of the work shares a similar basic experimental technique exploiting the advances in molecular genetics. Thus, even though the rest of this chapter will look at the literature in the different areas of genetics, this underlying unity must not be forgotten, and attempts will be made throughout to indicate overlaps. People in different branches of genetics probably never did work in isolation and they certainly do not now. In particular, geneticists who ignore the all-encompassing power of molecular genetics do so at their peril.

Short review (current awareness) journals

With the literature in genetics expanding beyond the grasp of even the most avid reader, and with so many libraries forced to prune their subscription lists, reasonably priced journals containing authoritative and well-referenced review articles of reasonable length are a godsend and have, for that reason, been given pride of place here. The monthly *Trends* journals have carved a vital niche for themselves, and most geneticists would agree that *Trends in Genetics* 1985– is invaluable. It provides a very readable collection of short reviews, expert comments on recent advances, technical tips and book reviews. The reviews are sufficiently well referenced to allow an in-depth literature study to be undertaken with ease. On the other hand, they are concise and written in such a way that they are ideal for both teachers and students. Articles of genetic interest also appear in *Trends in Biochemical Sciences* (TIBS), *Trends in Cell Biology, Trends in Biotechnology* and *Trends in Ecology and Evolution.*

A very useful series is *Current Opinion in Genetics and Development* (see p. 6). For example, two of the issues in 1994 deal with Genetics and Disease, and Oncogenes and Cell Proliferation. The articles are authoritatively written, well illustrated and well referenced. Another journal is *Current Biology (Dispatches from the Front Lines of Biology)*. These articles cover the full range of biological subjects, but enough deal with topics related to genetics to warrant a mention. These articles tend to be concise and provide useful summaries with key references.

Another valuable journal which covers all biological subjects is *BioEssays*, with many articles of interest to geneticists.

As most areas of genetics are changing at a daunting rate books date very quickly, and these current awareness journals are invaluable.

Although it follows in the traditions of its parent journal in publishing original papers covering the most recent advances in scientific research, *Nature Genetics* is an excellent journal for keeping abreast with general advances, in addition to the important original papers that it publishes.

Dictionaries

The value of dictionaries in a science as broad and complex as genetics is perhaps debatable, on the basis that if they are going to include almost everything then their coverage of any individual item must necessarily be very brief and superficial. However, because of the breadth of the subject it is often the case that someone working in one area can be unfamiliar with the terms, and even the concepts, used in other branches of the subject. In such circumstances even a rather brief entry in a dictionary may be just what is needed (see p. 39).

General textbooks

Because of the recent developments, particularly in the molecular area, textbooks written even a few years ago are rapidly out of date.

Several books covering most or all of biology contain much material on genetics that is very suitable for undergraduate use. One such book is *Molecular and Cellular Biology*, by S.J. Wolfe (Wadsworth, 1993), undoubtedly one of the most successful books to be published which contains much genetic material, largely but not exclusively with a molecular emphasis.

Most geneticists would probably agree that the textbook *An Introduction to Genetic Analysis* by A.J.F. Griffiths, J.H. Miller, D.T. Suzuki, R.C. Lewontin and W.M. Gilbert (5th edn., Freeman, 1993) has established the yardstick against which others should be measured. Its title is far too modest, as it is an authoritative survey of the whole subject and is ideal for courses at all levels.

A textbook which is less comprehensive but which nevertheless gives excellent coverage, with a good balance between classical and molecular aspects, is *Concepts of Genetics*, by W.S. Klug and

M.R. Cummings (3rd edn., Macmillan, 1991). The material is very well presented and profusely illustrated in colour. Alongside this should be mentioned *Genetics*, by D.L. Hartl (3rd edn., Jones and Bartlett, 1994). Earlier editions established it as a leader in the field and this latest edition is a first-rate text for the advanced student.

A new addition is *Understanding Genetics: a Molecular Approach*, by N.V. Rothwell (Wiley-Liss, 1993). This book takes an unashamedly modern approach by regarding the molecular aspects of genetics not simply as a part of the subject, but as the basis on which the whole edifice rests. This is not in any way to suggest that the coverage of classical genetics is inadequate, but molecular concepts are introduced early on and then used to underpin all that follows. This is a book that will appeal to a wide readership and could form the basic text for many genetics courses, particularly at an advanced level.

An important category of books on genetics that has burgeoned in recent years is directed towards the educated layperson who is scientifically literate but does not possess any substantial knowledge of the field. Richard Dawkins should take pride of place here, as his own personal and persuasive interpretation of genetics and a Darwinian world make compelling reading for the scientist and layman alike. *The Selfish Gene* (OUP, 1976) was followed by *The Extended Phenotype* (OUP, 1982) and, perhaps most famous of all, *The Blind Watchmaker* (Penguin, 1986). These ideas have been extended most recently in *River Out Of Eden* (Weidenfeld & Nicolson, 1995). All these books can be recommended for their informative content and illuminating style.

As a book for, say, an adult education class, *The Language of the Genes*, as S. Jones (Harper Collins, 1994) is ideal. The author makes considerable demands on his readers without overburdening them with too many facts or, worse, too much jargon. Another useful, but quite different, book in this category is *Essential Genetics*, as A. Hodson (Bloomsbury, 1994). The first half of the book is an introduction to genetics, seen largely from a historical perspective and full of intriguing insights. The rest is an alphabetically arranged encyclopaedia and dictionary. Although the book is of modest size and modestly priced, it contains a remarkable amount of useful information and well repays attention, even from those who may think they already know it all.

It would be fair to mention an 'anti-genetics' book, *Not In Our Genes*, by S. Rose, R.C. Lewontin and L.J. Kamin (Penguin, 1990), originally published in 1984. Its main message is that the euphoria following the technical triumphs of modern molecular biology has led to an unhealthily deterministic view of human biology, and that this has in turn had an excessive influence on politicians and other decision makers. Two of the authors are well known for their views on these matters, but this should not detract from the fact that they make a well reasoned case and question the conventional wisdom in a very thought-provoking way. Whether they are heretics or prophets crying in the wilderness will necessarily be a personal view.

Genetic maps and databases

Genetic maps are basic source material for geneticists, and published collections of the genetic maps of the most widely used organisms are invaluable. However, as so many of the most widely used organisms now have their own genome projects, all at different stages of development, these maps tend to be expanding at such a rate that they date rapidly, especially perhaps the human map, which has expanded from very modest beginnings to its present impressive state over two decades, and is continuing to grow exponentially. Thus the Human Genome Map published as a supplement by *Science* in 1990, although still a useful reference, is now very out of date. However, the *Genome Directory*, published as a supplement by *Nature* in September 1995, is the largest and most up to date compilation of sequence and mapping data on the human genome. The data represent a major early milestone in the progress of the Human Genome Project.

An updated compilation of genetic maps is published every few years by the Cold Spring Harbor Laboratory: *Genetic Maps: Locus Maps of Complex Genomes*, edited by S.J. O'Brien (6th edn., 1993; 7th edn. due in 1996). Genetic maps of particular species are published from time to time in several journals, particularly in microbiology where so many species are involved. An updated list of references to the genomes of bacteria is provided by the editors of *Microbiology*. Two recent examples are *Genetic Map of* Salmonella typhimurium, *Edition VIII*, by K.E. Sanderson *et al.*, *Microbiological Reviews* 1995; 59: 241–303 and a large series of papers on the Bacillus subtilis *Genome Sequencing Project* in *Microbiology* 1995; 141: 249–350. The recently published complete DNA sequence and genetic map of *Haemophilus influenzae* is discussed later under the heading of bacterial genetics. (see p. 148)

The value of hard-copy genetic maps and sequence collections has been to a very large extent overtaken by the production and ready availability of many genetic map and sequence databases. No fewer than 41 separate databases are listed in the September 1994 issue of *Nucleic Acids Research* (Vol 22 No 17, Database

Issue) and for each one a short and informative article is included. These give not only a useful summary of what the database contains and how it can be used, but also information about how to access it. The range of these databases is truly breathtaking, stretching from the vast ones such as GENBANK and the European Bioinformatics Institute (EBI) databases to specialized sequence databases such as *Database and Software for the Analysis of Mutations at the Human* p 53 *Gene*. In between these two extremes are databases dealing with individual organisms such as *Escherichia coli*, *Drosophila*, *Saccharomyces* and man (see also p. 145).

Another invaluable source is *Trends in Genetics: Genetic Nomenclature Guide Including Information on Genomic Databases*, published as a supplement with the March 1995 issue. Fifteen of the most popular organisms are covered in detail and, in addition to valuable guidance about nomenclature and how to get it right, there is a comprehensive guide to databases, culture collections and references to key articles, books and genetic maps. Information on how to contact these sources is provided, with locations, e-mail addresses and fax numbers. No practising geneticist should be without a copy of this.

Molecular genetics

There are a number of books and journals that deal primarily, even if not exclusively, with advances in molecular aspects of genetics, and it is important to note them separately, particularly as they include some of the most significant journals in biology. Key journals which are basic reading for all scientists in this area are *Proceedings of the National Academy of Science of the USA (PNAS)*, *Cell* and *Nature*. Many of the most noteworthy advances are first reported in these journals. Other key journals are *Gene*, *Molecular and General Genetics*, *Science*, *Journal of Molecular Biology*, *Nucleic Acids Research*, *Nature Genetics* and the new *Human Molecular Genetics*.

Among books pride of place must surely go to *The Molecular Biology of the Gene*, by James Watson (4th edn., Benjamin, 1986) and it is most appropriate that one of the founders of the subject should have written such an important textbook. Needless to say, it has expanded enormously from its classic first edition and has been divided into two volumes, but this is one of the indications that a book should be taken seriously in such a rapidly developing field. *Genes* was an excellent book when it first appeared and has been regularly updated, having now reached *Genes V* (B. Lewin, OUP, 1994). This is a superbly researched book and, like so many books of its type, it is beautifully produced and illustrated. However – and it is not alone in this – the paucity of references to the primary literature limits its value for many purposes including, unfortunately, as a sourcebook for students and teachers. It is clearly a matter for discussion and also for individual taste how thoroughly referenced a student textbook needs to be, but *Genes* seems to fall below an acceptable minimum in that it is generally impossible to trace the original sources. It is certainly very frustrating for both teacher and student not to be able to access an original reference. A book that takes the question of access to the primary literature very seriously and with excellent results, is one that might not, from its title, be taken as a serious contender in molecular genetics. *The Biochemistry of the Nucleic Acids* by J.N. Davidson (Methuen, 1950) started life as a pocket-sized monograph. The first seven editions continued to be written by Professor Davidson. The 11th edition, by R.L.P. Adams, J.T. Knowler and D.P. Leader (Chapman & Hall, 1992), covers much more than the title implies, and a wide range of topics in molecular genetics is dealt with in impressive detail. It is very thoroughly referenced with between 200 and 300 references to each chapter, and is a mine of information. It is copiously illustrated in black and white, and the absence of colour does not detract from its educational value. Although multicoloured artwork is aesthetically pleasing and, in some cases, can illustrate points in a way that black and white cannot, a student's memory tends to be monochrome, particularly when it comes to assimilating material for examination purposes.

The other category of book that should be discussed includes those that deal more specifically with recombinant DNA (genetic engineering), that is, with *in vitro* as opposed to *in vivo* aspects. One of the most important texts has for some time been *Principles of Gene Manipulation*, by R.W. Old and S.B. Primrose (4th edn., Blackwell, 1989). In addition to giving a very comprehensive account of the technical aspects of recombinant DNA, it also covers much of the molecular biology of the systems where the techniques are most profitably and frequently applied. It is readable and is very well referenced. A much less ambitious book, and very well suited for a beginner, is *An Introduction to Genetic Engineering*, by D. Nicholl (CUP, 1994), one of the Studies in Biology series. A book whose first edition (1983) dealt rather narrowly with the technical aspects of recombinant DNA, but whose second edition has been expanded to include an admirable coverage of the many fields of application, is *Recombinant DNA*, by J.D. Watson, M. Gilham, J. Witkowski and M. Zoller (2nd edn., Freeman, 1992). This is superbly written and presented and is an invaluable sourcebook, particularly at advanced undergraduate level.

Human and clinical genetics

The rate of progress in this field, not least as a result of the Human Genome Project and all the research activity associated with it, means that this is a particularly difficult area in which to keep abreast of recent developments. For this reason even the most recently published textbook will be seriously out of date. In addition, the sheer volume of information means that even the most comprehensive texts are going to do little more than scratch the surface, and will be of limited value as a reliable source of detailed information. This is particularly true of clinical genetics, where many individually rare diseases have to be covered, both from the point of view of symptoms, diagnosis and treatment, as well as the ever-increasing molecular detail that is accumulating. Having said that, however, there are many excellent books at all levels. The single most useful source of updated information is *Online Mendelian Inheritance in Man* (OMIM). This is an electronic version of *Mendelian Inheritance in Man*, by V. McKusick, first published by The Johns Hopkins Press in 1966 and regularly updated in many editions since. This has been online since 1985, but it is the recent ability to access it via the World Wide Web that has made the resource invaluable. A useful introduction to the subject was published in *Trends in Genetics*, July 1995 p. 291, and this article would be a useful starting point for anyone wishing to use OMIM. The system has been designed to make searching by subject as user-friendly as possible, and a system that permits Web browsing, such as Netscape or Mosaic, makes an astonishing amount of information available very rapidly. With very little practice search terms can be entered and information obtained in a way that is suitable for reading or, if required, printing. One very useful feature is that each entry gives the date when it was last updated. In the future, OMIM will be linked to other databases so that, for example, clinical aspects of a disease can be obtained in addition to the genetic information currently available. It is not possible to overestimate the importance of OMIM as a source of information on almost all aspects of human and medical genetics.

A clear explanation of the principles of the subject for use, for example, by medical students, is of primary importance. One of the classics is *Elements of Medical Genetics*, by A.E.H. Emery, now rechristened *Emery's Elements of Medical Genetics*, by R.F. Mueller and I.D. Young (9th edn., Churchill Livingstone, 1995);

this is an outstanding book for students. As with many books that have been through many editions, it does occasionally give the impression of having been updated rather than rewritten, but the incorporation of new material, particularly in the molecular area, is very well done and it is a worthy competitor for those books whose first editions have appeared more recently.

Essential Medical Genetics, by J.M. Connor and M.A. Ferguson-Smith (4th edn., Blackwell, 1993) is an excellent book for medical students and has the advantage of a first section on basic principles, followed by a second on clinical applications. This arrangement allows great flexibility and its approach is thoroughly up to date, with a strong molecular emphasis throughout. Two appendices on the basic theory of odds and probabilities are concise and well presented. A very useful companion volume is *Self-Assessment in Medical Genetics*, by J.M. Connor and J.R.W. Yates (Blackwell, 1986), which gives the reader a thorough test of their knowledge in a neatly organized way.

A book of limited scope but very well presented is *Molecular Medicine: an Introductory Text for Students* by R.J. Trent (Churchill Livingstone, 1993). It would be suitable for less ambitious courses for medical students. Another book in the same category is *ABC of Clinical Genetics*, by H.M. Kingston (BMJ, 1989), which is a useful primer.

There are other texts less specifically aimed at the medical student. *Human Heredity: Principles and Issues*, by M.R. Cummings (3rd edn., West, 1994) is an encyclopaedic treatment of the subject and is well written and profusely illustrated, with thought-provoking sections on issues of social or scientific concern. However, although the treatment of molecular techniques is adequate it is not written from a fundamentally molecular standpoint, a weakness for many readers.

Another book about which some of the same comments can be made is *Basic Human Genetics*, by E.J. Mange and A.P. Mange (Sinauer, 1994). It is very well written and is a mine of information presented in a basically classical manner. For a reader wishing to obtain a broad view of all the ramifications of the subject, however, this should be highly recommended.

An outstanding book written from an unashamedly molecular standpoint is *Molecular Genetics for the Clinician*, by D.J.H. Brock (CUP, 1993). Although of modest size and at a price that students can afford, it contains a remarkable amount of up-to-date information and would probably be the book of choice, not only for the curious clinician but also for undergraduate courses at a fairly advanced level.

The New Genetics and Clinical Practice, by D.J. Weatherall

(OUP, 1982) was a pioneering book when first published in that it recognized that the future of clinical genetics lay in the application of molecular techniques. Now in its third edition (1991), it is still a strong contender and is written with the authority of one who has been constantly at the forefront of research in the area.

An outstanding book that is encyclopaedic and based on the advances made possible by molecular biology is *Human Genetics: the Molecular Revolution*, by E.H. McConkey (Jones and Bartlett, 1993). It is well illustrated and copiously referenced and should be on the bookshelf of all serious students.

Another book about which many of the same favourable comments can be made is *Genes in Medicine: Molecular Biology and Human Genetic Disorders*, by I. Rasko and C.S. Downes (Chapman & Hall, 1995). This is full of up-to-date details, for example the nature of the gene for Huntington's disease, and is written in a refreshingly informal style. It would certainly make an excellent text for an advanced course and is reasonably priced.

A very useful little book on the molecular basis of human genetics is *The Human Genome*, by T. Strachan (Bios, 1992). The author has managed to include a remarkable amount of information and yet has retained a readable and informative style. It concludes with two particularly useful chapters on disease gene cloning and clinical perspectives.

A new book is *Human Molecular Genetics*, by T. Strachan and A.P. Read (Bios, 1995). Its coverage is comprehensive and brings the reader right up to date. It is well illustrated and written in a thought-provoking manner.

Because genetics impinges on many aspects of everyday life, it is vital to be able to condense the enormous complexity of modern genetics so that some of its practical importance, and even some of its intellectual excitement, may be conveyed to the lay reader. A highly successful attempt to achieve this is *Heredity and Your Family's Health*, by A. Milunsky (Johns Hopkins UP, 1992). This is a clear, readable and compassionate book which should be read by anybody who is involved with inherited disease. Although written for the layman, it contains much information and richly repays study by practising geneticists. Another book aimed at the same audience but on a rather less ambitious scale is *Our Genetic Future*, the report of a working party of the British Medical Association, published by OUP, 1992. It covers a wide range of issues connected with the application of modern genetics, but a substantial section is devoted to issues such as gene therapy, population screening for genetic disease and the Human Genome Project. A book that bravely attempts to introduce the beginner to the esoteric world of molecular genetics with great success is *Exons,*

Introns, and Talking Genes: the Science Behind the Human Genome Project, by C. Wills (OUP, 1992). It is a mine of useful information.

The genetics of cancer is important, and many of the books mentioned above deal with it in a thorough and up-to-date manner. Particularly worthy of mention are the relevant chapters in the books by Brock, by Rasko and Downes, and by McConkey, mentioned above. However, there are also books exclusively devoted to the genetics of cancer. *Advances in Understanding Genetic Changes in Cancer: Impact on Diagnosis and Treatment Decisions in the 1990s* is published for the Institute of Medicine by the National Academy Press, Washington DC, 1992. Although less than 100 pages long, it contains authoritative information on the genetic changes found in the different classes of tumour and concludes with a useful summary and a set of specific recommendations for future work. *A Practical Guide to Human Cancer Genetics*, by S.V. Hodgson and E.R. Maher (CUP, 1993) is more substantial and gives detailed accounts of a large number of tumour types where known genetic changes are involved. It is a book of great value to clinicians and researchers.

The journals that are central to human genetics fall into two categories. The first consists of mainstream journals dealing with advances in molecular genetics, such as *Nature*, *Cell* and *Science*. The second category consists of those that are exclusively devoted to human and medical genetics, such as *Journal of Medical Genetics*, *Clinical Genetics*, *Annals of Human Genetics*, *American Journal of Human Genetics* etc. Each of these journals has its own emphasis, particularly in the balance between clinical and research reports. In addition, many articles are published in journals devoted to disciplines such as haematology, cancer and general medicine, notably *The Lancet*. *Annual Review of Genetics* has authoritative articles on a wide range of human and other topics. Other titles in the *Annual Review* series are also useful, notably *Annual Review of Medicine* and *Annual Review of Physiology*.

Developmental and *Drosophila* genetics

Drosophila melanogaster, the humble fruit fly, has been one of the most important experimental animals in genetic research for many decades, and it has really come back into its own in recent years as a result of many elegant and informative studies on its developmental genetics. The study of the homeotic mutants at the molecular level led to the discovery of the 'homeobox', and then to the amazing discovery that similar genes appear to play comparable roles in the development of a wide variety of animals, including mammals. Another such motif found in *Drosophila* is the 'paired box', and this has also been found in mammalian genes which perform important roles in development, such as the *pax3* homologue that is mutant in Waardenburg's syndrome. For this reason developmental genetics and the genetics of *Drosophila* have been included in the same section. In October 1995, the Nobel Prize for Medicine and Physiology was awarded to geneticists for their work on the development of *Drosophila.*

Developmental genetics is not an easy field for the beginner and a book that makes the subject accessible to the non-specialist, and which is also fun to read, is worth its weight in gold. A strong contender is *The Making of a Fly: the Genetics of Animal Design*, by P.A. Lawrence (Blackwell, 1992). Written by an expert, this is a mine of information on *Drosophila* and its developmental genetics, but at the same time unfolds the complex processes involved in a way that makes it a pleasure to read. Progress from fertilized egg to adult animal is clearly a highly logical process, and this book presents it as such. It is a model of scientific writing and is to be highly recommended to anybody wishing to get to grips with this field.

The Genome of Drosophila melanogaster, by D.L. Lindsley and G.G. Zimm (Academic Press, 1992) is an encyclopaedic collection of data. *The Development of* Drosophila melanogaster, by M. Bate and A. Martinez Arias (Cold Spring Harbor Laboratory Press, 1993) is published in two volumes with a supplement, *Atlas of* Drosophila *Development*, which is not only informative but also beautifully presented in colour. At a more practical level, Drosophila: *a Laboratory Manual* and its companion volume Drosophila: *a Laboratory Handbook* (Cold Spring Harbor Laboratory Press, 1989) are invaluable guides to every aspect of *Drosophila* manipulation and experimentation.

As in any rapidly advancing field, up-to-date information can really only be obtained through a regularly updated publication in the form of a newsletter or by electronic means. The Drosophila *Information Service*, produced by the Department of Zoology, University of Oklahoma, provides a regularly updated series on the bibliography of *Drosophila* and its gene loci. An online version is FlyBase, which is accessible through the World Wide Web (p. 27, 144).

The standard text on development in its more general context is *Development*, by S.F. Gilbert (Sinauer, 1994). A rather smaller book, and now slightly out of date, is *Genetics and Development*, by J.H. Sang (Longman, 1984). This would serve as an excellent introductory text and merits inclusion here. There are substantial sections on animal development in several of the standard textbooks, in particular *Molecular Biology of the Cell*, by B. Alberts *et al.* (3rd edn., Garland, 1994) and *Molecular Cell Biology*, by H. Lodish *et al.* (3rd edn., Scientific American Books, 1995). The appropriate chapters in either of these books would serve as an excellent introduction. Both these books are essential reading for serious students of cell biology.

Of the journals dealing with developmental genetics, *Cell* carries many reports of new and significant findings, as do *Nature* and *PNAS*, as well as *EMBO Journal*, *Genes and Development*, *Development* and *Developmental Biology*.

An interesting footnote is provided by an article in *Trends in Genetics*, February 1995 p. 51, by V. Hartenstein, A. Lee and A.W. Toga, entitled 'A graphical digital database of *Drosophila* embryogenesis'. In effect, a cell-by-cell three-dimensional digital map of the embryo is generated from confocal sections, and data, for example on gene expression, can then be incorporated and displayed in any way desired. When this database is completed and made available through the World Wide Web, it will clearly be invaluable for *Drosophila* biologists.

Population and evolutionary genetics

Geneticists tend to be divided into two categories with respect to population genetics. The first are specialists whose command of the mathematical aspects of the subject sometimes makes their work impenetrable to the majority. The other group find the rigorous mathematical treatment beyond them, and are therefore forced to understand the subject at a somewhat superficial level. For human geneticists the monumental *The Genetics of Human Populations*, by L.L. Cavalli-Sforza and W.F. Bodmer (Freeman, 1971) is still a classic text. For those who find this volume somewhat daunting, a later book by the same authors, *Genetics, Evolution and Man* (Freeman, 1976) is a highly readable account of the subject, presented with the necessary mathematics to make the results comprehensible without being intimidating. It is a mine of interesting and useful information and a model of scientific writing.

As an up-to-date and advanced treatise, *Principles of Population Genetics*, by D.L. Hartl and A.G. Clark (2nd edn., Sinauer, 1989) is probably the text of choice. For the enthusiastic beginner a very good introduction is *A Primer of Population Genetics*, by D.L. Hartl (2nd edn., Sinauer, 1988). More evolutionary aspects are well covered in *Evolutionary Biology*, by D.J. Futuyama (2nd edn.,

Sinauer, 1986) and *Evolution*, by M. Ridley (Blackwell, 1993). An outstanding book by one of the leaders is *Evolutionary Genetics*, by John Maynard Smith (OUP, 1989). The reader is helped through some of those areas that are often found difficult and the style of writing makes the subject approachable. No list of books in this field would be complete without mention of the four volumes by Sewall Wright, one of the founding fathers of the subject, *Evolution and the Genetics of Natural Populations* (Chicago UP, 1968–78).

Genetic analysis at the population level is now being approached more and more by molecular methods, and the widespread application of DNA fingerprinting is one important example of this. A good background to modern developments, but clearly itself rather out of date, is *Evolution of Genes and Proteins*, by M. Nei and R.K. Koehn (Blackwell, 1983). A very useful volume is *Molecular Genetic Analysis of Populations*, edited by A.R. Hoelzel (IRL, 1992), one of the Practical Approach series. Although this book is primarily devoted to practical protocols, each chapter has a useful theoretical introduction. A short but informative volume is *Molecular Genetic Ecology*, by A.R. Hoelzel and G.A. Dover (IRL, 1991) in the In Focus series, which gives a good grounding in less than 100 pages.

The Cambridge Encyclopaedia of Human Evolution, edited by S. Jones, R. Martin and D. Pilbeam (Cambridge University Press, 1992) gives an impressively comprehensive overview.

Research articles on population and evolutionary genetics tend to be widely scattered, but there are nevertheless a few places where such work is concentrated. Key journals are *Heredity*, *Evolution*, *Journal of Evolutionary Biology*, *Genetical Research* and *Journal of Heredity*. In addition to *Annual Review of Genetics*, *Annual Review of Ecology and Systematics* contains many useful articles. Of the current awareness journals, in addition to *Trends in Genetics*, *Trends in Ecology and Evolution* is particularly useful.

Microbial genetics

Bacteria

The bible of classical bacterial genetics, written by one of the founders of the subject and published just a short time before the onslaught of the molecular revolution, is undoubtedly *The Genetics of Bacteria and Their Viruses*, by W. Hayes (2nd edn., Blackwell, 1968). For a lucid explanation of the reasoning behind the fundamental principles and their experimental basis one cannot do better than dip into this book, despite its age. It does, of course, come from an era when bacterial genetics was more or less synonymous with the genetics of *Escherichia coli* and its close relatives, such as *Salmonella*. It is important to appreciate that this is no longer the case and, even if *E. coli* is in some respects still the leader, there are now a number of other species in contention, particularly when it comes to total sequencing projects.

A useful book is *Genetic Elements in* Escherichia coli, by P. Smith-Keary (Macmillan, 1988). Complex subjects such as the genetic regulation in phage λ, and the mechanism of genetic recombination, are treated in a way that makes this book particularly suitable for students.

Perhaps the best introduction to bacterial genetics is *Molecular Genetics of Bacteria*, by J.W. Dale (2nd edn., Wiley, 1994). This new edition takes a thoroughly molecular approach and provides an excellent grounding to the genetic systems found in bacteria. Reasonably priced, it should interest students in many related biological disciplines. *Modern Microbial Genetics*, edited by U.N. Streips and R.E. Yasbin (Wiley-Liss, 1991) is an encyclopaedic multiauthor work which covers the whole field of microbial genetics, both prokaryotic and eukaryotic, in a systematic way. A group of chapters on basic molecular biology is followed by a thorough treatment of *E. coli* genetic systems and a chapter of bacteriophage genetics, which is largely, however, confined to a consideration of phage T4. Chapters on yeast, *Agrobacterium* and *Aspergillus nidulans* and other filamentous fungi give useful summaries. A final very interesting chapter outlines possible approaches to the genetic analysis of those species of bacteria (still, after all, the majority) that do not possess known genetic systems. This is undoubtedly a book for the serious student.

Most of the major textbooks of microbiology have sections on bacterial genetics, although in some cases this may not amount to more than a single chapter. One of the best-known microbiology texts, *Microbiology*, by B.D. Davis, R. Dulbecco, H.N. Eisen and H.S. Ginsburg (4th edn., Lippincott, 1990), is an important exception and contains a very substantial section on bacterial physiology and genetics. This book undoubtedly deserves its reputation as the reference work of choice, and is well referenced to supplementary reading material.

Detailed genetic maps and DNA sequences from individual bacterial species are published in a number of journals. The SGM provides a list of references of what may be regarded as standard maps for a number of the most popular species, and this list is updated. A landmark publication is the complete genetic map and DNA sequence of *Haemophilus influenzae* (*Science* 1995; 269:

496–512). This was the first complete DNA sequence of a free-living organism and, at 1 830 137 base pairs in length, it was the largest DNA molecule sequenced to date. If *E. coli* had temporarily been knocked off its pedestal, at least its genome is twice as large. That of man is nearly 2000 times as large!

Many of the mainstream microbiology journals devote significant sections to bacterial genetics, such as *Microbiology* and *Journal of Bacteriology*. Others are more exclusively dedicated to the subject, such as *Molecular and General Genetics* and *Molecular Microbiology*. *Infection and Immunity* also publishes many papers of genetic interest. Of the review journals, in addition to *Trends in Genetics*, *Trends in Biotechnology* carries useful reviews on the more applied aspects of microbial genetics. For more substantial reviews *Microbiological Reviews*, *Annual Review of Microbiology* and *Advances in Microbial Physiology* are key sources.

Viruses

Many of the most useful review articles appear in journals such as *Trends in Genetics*, *Annual Review of Genetics* and *Advances in Virus Research*. *Journal of General Virology* and *Journal of Virology* are key journals for research papers. Many of the most noteworthy advances in this area are inevitably reported in the mainstream molecular biology journals such as *Nature*, *Science* and *Cell*.

Yeasts and fungi

The bible of classical fungal genetics, comparable to the book on bacterial genetics by Hayes mentioned above, is *Fungal Genetics*, by J.R.S. Fincham, P.R. Day and A. Radford (4th edn., Blackwell, 1979). It is a comprehensive account of the genetics of both yeasts and filamentous fungi and, despite its age, is still an invaluable sourcebook. A very recent book that covers advances in the physiology, genetics and differentiation of the filamentous fungi in impressive detail is *The Growing Fungus*, edited by N.A.R. Gow and G.M. Gadd (Chapman & Hall, 1994). Extensive coverage of the genetics of both yeasts and filamentous fungi is given in *Gene Manipulation in Fungi*, edited by J.W. Bennett and L.L. Lasure (Academic Press, 1985), and, on a more modest scale, in *Gene Structure in Eukaryotic Microbes*, edited by J.R. Kinghorn (IRL, 1987).

Owing to the importance of these microorganisms in biotechnology, there are many books devoted to the genetics of those species of particular commercial importance. An important example is *Applied Molecular Genetics of Filamentous Fungi*, edited by J.R.

Kinghorn and G. Turner (Blackie, 1992). As well as comprehensive treatment of the more traditional experimental species, there are chapters on the cultivated mushroom *Agaricus* and the many different types of oriental mushrooms. In this category come *Genetics and Molecular Biology of Industrial Microorganisms*, edited by C.L. Hershberger, S.W. Queener and G. Hegeman (ASM, 1989) and *Molecular Industrial Mycology: Systems and Applications for Filamentous Fungi*, edited by S.A. Leong and R.M. Berka, (Dekker, 1991). Two important books are Aspergillus: *50 Years On*, edited by S.D. Martinelli and J.R. Kinghorn (Elsevier, 1994) and *The Genus* Aspergillus *From Taxonomy and Genetics to Industrial Application*, edited by K.A. Powell, A. Renwick and J.F. Peberdy (Plenum 1994).

Research papers in fungal genetics are published in journals already mentioned, such as *Molecular and General Genetics*, *Molecular Microbiology* and *Yeast*.

Plant genetics

The classic text on plant breeding is still *Principles of Plant Breeding*, by R.W. Allard (Wiley, 1960). A recent addition is *Plant Breeding: Principles and Prospects*, edited by M.D. Hayward, N.O. Bosemark and I. Romagosa (Chapman & Hall, 1993), which covers all major aspects of plant breeding and discusses the prospects that the new technologies may offer in the future. A useful text on classical genetics is *The Genetics of Flowering Plants*, by V. Grant (Columbia UP, 1975).

With the advent of molecular genetics plant genetics has, like every other area, been largely revolutionized. The discovery, for example, of the *Agrobacterium* transformation system has allowed direct manipulation of plant cells and the production of transgenic plants. A volume devoted to this topic and dealing with both plants and animals is *Transgenic Modification of Germline and Somatic Cells*, edited by R.B. Flavell and R.B. Heap (Chapman & Hall, 1993). A useful primer, one of the Practical Approach series, is *Plant Molecular Biology*, edited by C.H. Shaw (IRL, 1988). An important series is *Oxford Surveys of Plant Molecular and Cell Biology*, edited by B.J. Miflin (OUP, 1984–.). A useful laboratory manual is *Plant Molecular Biology Manual*, edited by S.B. Gelvin, R.A. Schilperoort and D.P.S. Verma (Kluwer, 1988), which is regularly updated with new sections.

Research papers are widely distributed within the literature of botany, general genetics and molecular biology. Among the botanical journals, *The Plant Cell*, *Plant Molecular Biology* and *Plant*

Physiology all publish papers of genetic interest. *Plant Physiology* carries a section specifically devoted to gene regulation and molecular genetics, and also a plant gene register. Many papers on plant genetics also appear in *Theoretical and Applied Genetics* and *Heredity*.

CHAPTER SIXTEEN

Biotechnology and patents

H.V. WYATT

Today we are all molecular biologists. (S. Brenner, 1989)

The techniques of molecular biology have given rise to the biotechnology industry and the renaming of many activities. Research has been dominated by hopes of economic exploitation and the search for money. Patents and the hope of patents have become very important and threaten the sharing of research results. One aspect has been the proliferation of biotechnology newsletters bringing selections from the research, production and financial worlds. The more financially oriented, the more expensive the newsletter. A useful look at the many newsletters is obtained by browsing through *BioCommerce Abstracts* (see below). The list of serials currently taken by or accessed by BLDSC has 43 titles beginning *Biotech* . . . and another seven also with *Biotec* . . . or a variant such as *Bioprocess*: a few are journals, but many are newsletters. Some titles seem to have disappeared and some are not taken by BLDSC. *Biotechnology Bulletin* monthly from the UK is mainly for businesses. *European Biotechnology Newsletter*, about eight pages every two weeks, is again mainly for businesses and has short notes on a few patents, but is interesting to read. From the same office come *Biofutur*, *Flash Etats-Unis* and *Flash Japon*, all monthly and with articles mainly in French.

Applied Genetics News monthly from the USA is also available on line and via the Internet. It has news from industry and many reports of new findings, products etc., but with no citations to the literature these would be difficult to follow up. However, there are useful short summaries as introductions to items. *Genetic Technology News*, monthly from the USA, is mainly company news with some details of new US patents. There are references and addresses and telephone contacts. There is a good short review of some potential market in each issue. *Genetic Engineering News*, every two weeks, USA, is a product newspaper which is free except for a postage charge, modest for the USA but expensive elsewhere. There are short articles, reports of meetings and news of future meetings, but the bulk is product advertisements and company news.

The monthly *Antimicrobics and Infectious Diseases Newsletter*, which was formerly *Antimicrobic Newsletter*, is concerned with new and anticipated changes in regulatory standards and guidelines for antimicrobial agents. It reports new drugs and *in vitro* diagnostics and discusses diagnosis and treatment of infectious diseases.

Agrow World Crop Protection News, with offices in the UK, USA and Japan, appears every two weeks with 20 pages. It is an expensive business newsletter, but has news of legislative developments and new products. It sells a series of *Agrow Reports*, which are very expensive: e.g. *Adjuvants in Crop Protection*, *Post-Emergence Herbicides: a Market in Perspective*, *The Indian Agrochemical Market* (see also p. 101).

Almost all disciplines use the new technique of biotechnology. Publishers one might have previously thought cautious include biotechnology in their lists. CAB International, for example, publishes many biotechnological books and manuals on plants, animals and parasites and pests.

The monthly *Bio/Technology*, now called *Nature Biotechnology*, links scientists and industry, and its parent *Nature* organizes conferences in the major biotechnology countries. The Commentary and News Analysis sections are similar to the News and Views sections in *Nature* and equally important. Apart from highly technical scientific papers, there are also more general articles on patents etc. One special section included an international directory to regional development organizations. *Europroduct Focus*, distributed at selected meetings, is a product magazine with a few articles and abstracts of papers published in *Bio/Technology*. There is an annual, e.g. *Bio/Technology's 1996 International Buyers' Guide*, which has lists of products and suppliers as well as informative comparative tables for antibodies and chromatography. The *EFB Newsletter* of the European Federation of Biotechnology, also published by Macmillan, has news of meetings. Most importantly, it has details of publications, abstracts, reports etc. from bodies such as the OECD, and the grey literature (see SIGLE p. 32). **SEC: Healthcare, Pharmaceuticals and Biotech** is a CD-

ROM with full-text company information on 360 companies as reported to the US Securities and Exchange Commission.

The USA has important regulatory bodies for certain products: the FDA and USDA in particular are the premier agencies which dictate whether a product is to be licensed for use. The European Union is also important. Knowledge of and compliance with US requirements is vital at all stages of development, and there are many training sessions given, both in the USA and in Holland. An alternative to attending these expensive workshops is to buy videotapes from suppliers such as BioConferences International, Maryland 20814–4536, USA. Examples are: FDA consensus workshop on preclinical safety testing of monoclonal antibodies; European regulatory policy developments; and Safety and regulatory issues in transgenic production of biopharmaceuticals. There are also videotapes on good clinical practice, biotherapies of human genes, antisense and vaccine therapy, and video guides to targeted mutagenesis and transgenic techniques in mice.

Biotechnology seems to have spawned more than its fair share of multi-authored, multivolume series. *Biotechnology*, from VCH, is described as 'a whole science in 12 volumes'. The second edition began in 1991 with volume 4, with 12 published by the end of 1996.

Secondary serials

There are a number of abstracting serials, some of which are subsets of larger databases. The various journals, newsletters and secondary serials, with their limited variations of 'bio' in the title, are very confusing, especially as the CD-ROM catalogues give no indication of the parent bodies.

BioCommerce Abstracts and Directory contains 300 000 press reports 1981– from newsletters, newspapers, magazines and science journals: it is a business database for those interested in clients, markets etc. The directory contains information on 2000 organizations worldwide. The database may be searched online. The abstracts on CD-ROM are issued quarterly. *Abstracts in BioCommerce* 1978– appears every two weeks in print, on floppy disk and as CD-ROM, with very short summaries of 1000 very recent articles. A subset is issued as *BioCommerce Financial Abstracts*. *Biotech Knowledge Sources* 1988–, monthly, is a very good source of information about conferences, market surveys, books etc. There are short summaries for each. There are also feature articles and a useful list of publishers, including major distributors in the USA and Europe, with addresses, fax numbers etc. However, an address of 'New York, USA' for Garland Publishing is not helpful. There is an annual index.

Current Biotechnology Abstracts 1983– is produced from core journals and newsletters with patents. It is only about a third of the size of *Biotechnology Abstracts* (see below), but has better news coverage. It is available online.

Current Advances in Applied Microbiology and Biotechnology is part of the CABS database (p. 77).

BioBusiness is a BIOSIS database with 700 000 items 1985– for pharmacology, biotechnology, agriculture, food and pollution. In addition to the sources used for the other services, newspapers, business and trade publications are also used. BIOBUSINESS/RN is the same database but may be searched using CAS Registry numbers for specific chemicals. BIOSIS also produces FOOD INTELLIGENCE ON CD 1985–, with a total of 50 000 items each year, including 20 000 scientific references and 1000 patents culled from the BIOSIS databank, which includes 340 business and trade publications (see also p. 76).

The Royal Society of Chemistry and DECHEMA produce *Chemical Engineering and Biotechnology Abstracts (CEABA)* and from its database *Current Biotechnology Abstracts (CBA)* and *Biotechnology: Apparatus, Plant and Equipment*. They are available both online and in print.

Biotechnology Abstracts 1982– covers 1200 journals and all aspects, from gene manipulation to waste disposal and purification. There is a CD-ROM which is updated quarterly. **Biotechnology and Bioengineering** is a CD-ROM 1984– with 500 000 abstracts.

Patents

Patents are an important source of information and are becoming more so as competition for research funding becomes more intense. All institutions now required researchers to patent discoveries, even though this may delay publication in the scientific literature (see p. 00). Researchers should not attempt to take out a patent without expert advice. All patents are given an International Patent Classification (IPC) code and the *World Patents Index* 1974–, covering 30 countries, can be searched using the IPC code. Recent changes in the IPC have made it easier to search for biotechnology items such as monoclonal antibodies. Protection under patent law differs according to country, e.g. plant and animal varieties are not protected in India, China, Brazil and Mexico, but are in Japan, the USA and Australia. Microorganisms are not protected in India, Mexico, Brazil and Canada, but are in most other important countries (see p. 149). Many patents are now published with their search report, so one can trace the development of an idea or do a forward search, rather like using *Science Citation Index*. So far this is only possible with USA, European and a few other patents.

Patents. A Basic Guide to Patenting in Biotechnology, by R.S. Crespi (CUP, 1988) includes European and Japanese Patent Office guidelines. There are sections on copyright for DNA and legal protection for animals and plants. However, there have been new rulings about transgenic mice, for example, in 1990, and later by the European office, and the patenting of DNA sequences is still in flux. *Introduction to Patents Information*, edited by S. van Dulkan (2nd edn., BL SRIS, 1993) is a basic guide.

In 1980, the US Supreme Court ruled that a microorganism could be patented, and since then both animals and plants have been patented: for example, the Harvard oncomouse, a transgenic animal, in the USA in 1988 and in Europe in 1992. These legal opinions have created fierce opposition and anger in third-world countries. *Animal Patents. The Legal, Economic and Social Issues*, edited by W.H. Lesser (Macmillan, 1990) discusses many of the issues.

United States Patents and Fungal Cultures, by S. Jong and J.M. Birmingham, in *MIRCEN Journal* 1989; 5: 411–450, lists nearly 300 patents by patent reference, fungi available from ATCC (see p. 150), and classifies by use.

CASSIS is a CD-ROM database of five million records, including all the US patent documents 1790– . The 1.5 million US patents from 1969 are indexed by the patent holder's name and by the descriptive title, and many also have a synopsis of the patent. Discs are updated every two months. Copies of searches can be made on to floppy disks. The Automated Patent Searching (APS) database, which contains all US patents 1989– , is updated monthly.

FIRST and ESPACE, from the European Patent Office, updated weekly and with full texts, may be searched by name and title. However, if the name of the applicant is known it is quicker to use the European Patent Office (EPO) weekly paper journal. EPO is becoming very important as it accepts patents for all European Countries.

The Science Reference and Information Service (SRIS) of the British Library London, (see p. 98) provides priced services for business, environmental, biotechnology, Japanese and science and technology information.

Currentscan is a current awareness service for companies wishing to know about new patents. CD-ROM databases at SRIS are available for searching, printouts and copying on to floppy disks (which must be purchased). SRIS holds more than 33 million patents from all over the world – the largest collection anywhere – and supplies photocopies by PATENT EXPRESS. There is a free *Patents Information News*. Existing translations of patents will be found and professional translators will make a translation of a patent if required. Searches of chemical structures can be made. More than a million patents are online and any one can be found in two minutes from the more than 1000 CD-ROM discs, answering about 15% of the annual requests. Eventually it is hoped that copies may be ordered by touch-tone telephone and delivered automatically by fax, or perhaps via the Internet. The British Patent Office, London, publishes a free booklet, *Patents: a Source of Technical Information.*

There is a UK Patents Information Network available at 13 libraries: Leeds Public Library in England has more than 18 million patents. All libraries in the UK and many in Europe now have all new patents on ESPACE CD-ROM, rather than as paper documents. The research libraries of the New York Public Library have all US patents and a great many from other countries. Public libraries in other countries may also have patent collections.

Chemical Abstracts and *Biotechnology Abstracts* contain relevant patent abstracts, as do a number of biotechnology journals. *Vaccine*, now online as well as in print, has details of patents, but photocopying of the journal is forbidden. *Biologicals*, the journal of the International Association of Biological Standardization, carries patent and licence reports as well as quality control etc. There is a small patent section in *Oceanographic Literature Review Deep-Sea Research, Part B*, a Pergamon journal which uses the company PATSEARCH program. *Biotechnology Advances* 1983– has reviews and patent abstracts.

International Industrial Biotechnology 1981– is aimed at the scientist, with short review articles, information about firms, equipment, patents, books, meetings etc.

Derwent Publications process 13 000 patent documents each week, producing an informative abstract for each and relating it to previous patents where appropriate. *Chemical Patents Index*, weekly, has sections on pharmaceuticals, agricultural chemicals and food, and *World Patents Abstracts* has sections on agriculture and health. *Biotechnology Abstracts* 1982–, every two weeks, publishes 15 000 abstracts a year from more than 1300 journals and conference proceedings, as well as patent documents. A CD-ROM is updated quarterly and has more than 150 000 records from 1982–, including text, formulae and structures. The *World Patents Index*, with more than seven million patents, and

Biotechnology Abstracts may be accessed online. Derwent services are comprehensive and mainly suitable for industry or very large laboratories: Ringdoc provides 1000 pharmaceutical abstracts a week, and Vetdoc and Pestdoc cover veterinary products and pest control. Ringdoc is available on CD-ROM. More details are given in the free *Derwent Information* newsletter.

Derwent has a unique database, GENESEQ, which contains nucleic acid, protein (actually polypeptide) and probe sequences from patent applications and granted patents. The database stores all such sequences from 1981 and may be searched using the usual programs. It may be used in conjunction with IntelliGenetics to search the other sequence databanks, GenBank, EMBL, PIR and SWISS-PROT. GENESEQ contains small polypeptides and peptides which may not be in the other databanks, as well as sequences in patents which may not appear in the literature or the databanks. There is cross-referencing to both *WPI* and *Biotechnology Abstracts*. In 1995 GENSEQ contained more than 110 000 sequences and was adding 30 000 more each year: it claimed that more than half were not found in any public sequence database.

Also from Derwent is the Gene Therapy Database on CD-ROM 1982–, updated monthly, which seems to be a subset from *Biotechnology Abstracts*; the sections on gene transfer systems and vectors and gene manipulation will be of interest to many molecular biologists and virologists. Eight articles and patents are selected each month for reviews in which the findings are interpreted. Abstracts are available. *Gene Therapy Newsletter* contains reviews of key papers and patents, news items, and is a forum for legal and ethical debate. About 150 literature sources and 50 citations are given in each issue. The academic subscription rate is considerably reduced. This is the first of a new series of Bioscience Innovations current awareness databases, launched in 1995. The second, Environmental Biotechnology database 1982–, is similar, with a very wide interest base, e.g. agrochemicals, biocatalysis, fermentation waste disposal, nucleic acids etc. Both databases have backfile CD-ROMs from 1982, but diskettes from 1995– only.

Derwent also has a series of CD-ROMs on agrochemicals: Derwent Crop Protection 1968– covers pesticides, plant and insect growth regulators, attractants, repellants and biological control. Information comes from 1200 journals, with 40% of the abstracts coming from conference etc. literature. There is a printed version. A subfile and CD-ROM is the Derwent Crop Protection Registry of 8000 chemicals which have been indexed at least twice in the main file. Chemical activities, structure etc. are given.

Biotech PatentSearch 1980– is a CD-ROM, updated monthly,

from Microinfo, a firm which produces a host of specialist patent CD-ROMs.

Russia and other eastern European countries, such as Poland, are keen to offer facilities and services in universities and commercial units: directories of services etc. are published and advice may be obtained through embassies. An example is *Who's Who in Biotechnology in Russia and the Republics*, from Technology Detail, York YO3 6JH, UK. There is a regular column on biotechnology in eastern Europe in *Microbiology Europe*. Recently Russian microbiologists advertised their services for contract research in *Nature*. The very large Institute of Bioorganic Chemistry in Moscow cost US$ 150 million and contains a vast biotechnology pilot production plant. The International Institute of Biotechnology, Canterbury, UK, provides a forum for education and information as well as research initiatives in microbial biodiversity in tropical ecosystems, biodiversity for biotechnology innovation and marine biotechnology (including coral reefs and the Deep Sea Project on microorganisms). There is *Biotechnology Central/Eastern Europe Newsletter* 1995–.

The OECD has a Paris office for biotechnology and is active in safety and other important subjects. The first workshop was published as volume 84 of *Developments in Biological Standardization* 1995, *Non-Target Effects of Live Vaccines*.

In the USA the National Center for Biotechnology Information at the National Library of Medicine is mainly concerned with the development of databases, their use, and research into analysis of the data. It is responsible for the GeneBank DNA sequence database. Data are exchanged with EMBL and the DNA Database of Japan (DDBJ), and material is received from the National Agricultural Library (NAL) and the US Patent and Trademark Office. The Center provides the Directory of Biotechnology Information Resources (DBIR) database online: about 30% of listings are international. The Institute for Biotechnology Information is at Raleigh, Carolina 27709–4569, USA.

Directories

The International Biotechnology Directory 1996, by J. Coombs and Y.R. Alston, annual (MacMillan) is a guide to 10 000 companies, centres etc. in 21 countries, of which 2500 entries are for North America. There is a buyer's guide to services, products and research. The entries are minimal and not very informative: *Linscotts Directory of Immunological and Biological Reagents* would probably be more useful because it lists all the hundreds of *hybrid-*

omas, who markets them and suppliers' addresses in the scientist's country or area (see p. 44). *Biotechnology Guide, USA Companies, Data and Analysis*, by M. Dibner (3rd edn.,) Institute for Biotechnology Information 1994 appears every two years. *The UK Biotechnology Handbook '96*, (7th edn., annual) from BioCommerce Data Ltd, has entries for over 700 organizations with details of senior staff, products etc. See also *BioCommerce Abstracts and Directory*, above. *Nordic Companies in Biotechnology*, distributed by BioCommerce Data (1994) lists about 250 Scandinavian companies. *The Pharmaceutical Companies Fact File 1995* (4 vols, 2nd edn., Scrip Reports) is available in print or as a CD-ROM and contains complete details of over 750 research-based pharmaceutical and biotechnology firms.

Biotechnology Information (Germany) or BIKE 1987–, in German, is an online database of all aspects of biotechnology, including consultants and profiles of organizations. Israel Biotechnology Database is a new CD-ROM directory.

In Europe, a series of 33 open learning books on biotechnology is supplemented by seven interactive computer programs (Open Universiteit of The Netherlands and University of Greenwich, UK).

Ethics and safety

There has been intense public interest in biotechnology, with many groups opposed to some of the more controversial aspects and experiments. *GenEthics News* 1994– covers genetic engineering, ethics and the environment in six issues a year, each of about 12 pages. Each issue tables details of releases into the environment of genetically engineered organisms around Britain, and provides information on a whole range of interesting topics. *Genetically Modified Organisms: a Guide to Biosafety*, edited by G.T. Tzotzos (CABI, 1995) is a comprehensive review of safety as seen by a UN agency for development. The UK Department of the Environment has commissioned a series of *Genetically Modified Organisms Research Reports* 1994–. The first was *Genetically Modified Crops and Their Wild Relatives – a UK Perspective* (1994). The second, which was actually published first, *Genetic Modification of Fish – a UK Perspective* (1994). The report includes a directory of groups and their interests (the pagination is peculiar). The reports were actually made earlier and there was a delay in their publication, but an addendum includes 1993 references. Both are good reviews with useful bibliographies and are, of course, included in the *British Reports, Translations and Theses*

(BLDSC) (see p. 109). There is a genetically modified organisms – database at the University of Birmingham, UK. The Hastings Center, New York, discusses difficult moral issues in our society and publishes the *Hastings Center Report*, a supplement of which considered the 'Brave new world' of animal biotechnology (January–February, 1994).

Databases in molecular biology and genetics

Occasional listings of databases are made in journals but rapidly become obsolete. Particular listings have been:

genomic databases, in *Science* 1991; 254: 202–207, useful for descriptions of each database;
the special issue of *Nucleic Acids Research* 1994; 22 part 17 (September) 3441–3665 (see below). The journal is on line from 1996 with a three-year backfile;
a series in *The Biochemist* 1994–1995;
LIMB is the listing of molecular biology databases from Los Alamos National Laboratory (see also p. 104 for India).

Two inexpensive CD-ROMs are Molecular Biology Database Collection and the Molecular Biology Software Collection, both from the European Bioinformatics Institute. An expensive CD-ROM from John Libby Eurotext is Genome Interactive Databases, with associated references and maps.

The main databases publish newsletters about their activities, e.g. the quarterly *Protein Data Bank* (PDB) and *NCBI News* three times a year. Progress is being made on the software which will enable the various databases to be integrated. The different cDNA cloning and sequencing projects should be integrated with the Human Genome Mapping Project, with the gene products and chemical structure databases. The Cambridge Crystallographic Data Centre, CCDC, has 2D and 3D details of nearly 150 000 small molecules. The scale of the information being received is indicated by the doubling time of about 18 months for entries in the Protein Data Bank and 1000 entries a day of the human cDNA ESTs (see below). The databases, researchers and others have produced software packages for searching and comparing sequences. Some firms will do the searching for customers. IntelliGenetics provides the BIONET Online service using 16 sequence analysis programs to search 19 databanks. The company brochure provides a compact description of these databanks, including three maintained by IntelliGenetics: VectorBank, of frequently used cloning vectors; NASITE, of nucleic acid patterns and motifs; and KeyTool, of sequence patterns.

Sequences have been conserved throughout evolution although with different functions. This research requires that the different databases should be compatible. The databases collaborate, so that since 1987 GenBank, EMBO and DDBJ (see below) have pooled their data. GENINFO, at the National Library of Medicine (NLM) at Bethesda, Maryland, will store all published articles on biotechnology, including those from MEDLINE, and act as a quid pro quo database of unpublished nucleotide and peptide sequences.

Mammalian DNA

The Online Mendelian Inheritance in Man database (OMIM) is based on *Mendelian Inheritance in Man*, by V. McKusick; it includes clinical data and is cross-referenced to the Human Genome database (GDB), which is a reference library for the information held in other databases. Both are at Johns Hopkins Medical Institutions in Baltimore. The Mammalian Gene database is now at the European Bioinformatics Institute (EBI) near Cambridge, UK.

GenBank DNA Sequence databank, which is updated eight times a year, was at Los Alamos but is now merged with the EMBL Nucleotide Sequence Data Library and is operated by the NLM. Plant and patent sequence data are obtained from the National Agricultural Library and the US Patent and Trademark Office. In early 1995 GenBank was receiving 1000 human cDNA expressed sequence tags (ESTs) per day from the Genome Sequencing Center in St Louis: there were already nearly 100 000 ESTs in the database. GenBank may be accessed by e-mail, Internet, World Wide Web etc., and from CD-ROMs. The Internet access has advantages over the CD-ROM version: there are daily updates, three-dimensional structures may be viewed, and the sequence-related subset of MEDLINE is five times the size.

The EMBL database and the MRC Genomic Resource Center have both moved to the Sanger Center at Hinxton near Cambridge, UK. The DNA Database of Japan (DDBJ) is at the National Institute of Genetics, Michima. GENATLAS is a catalogue of mapped genes and other markers at Hospital Necker, Paris, France. The Genomic Interactive Databases (GID) is a CD-ROM which is updated twice a year. It combines a number of databases, with over 3500 genes and the location of 800 genetic disorders and 2000 markers. Another database contains information on YACs and sequence-tagged sites etc. More than 20 000 references are included. It is intended to be a powerful tool for use with the Human Genome Project (see *Biotech Knowledge Sources* 1995; 8 (4): 3–4).

The Mouse genome mapping database, GBASE, is at the Jackson Laboratory, Bar Harbor, Maine 04609 together with the Homology Database and Project of genetic data on humans, mice and 23 other species. The Induced Mutant Resource (IMR) maintains and distributes targeted mutant and transgenic mice. Over 150 strains have been obtained since 1992 and about eight are added each month. There are lists, and new strains are reported in *JAX Notes* or by e-mail or on the World Wide Web. Over 1700 strains of mice are held at the laboratory (see also p. 155). A European mouse gene repository and major genetics centre is being established at Monterotondo near Rome for 1997 as the European Mutant Mouse Archive. *Mouse Genetics. Concepts* and *Applications* by L.M. Silver (OUP, 1995) is a methods book of tools and procedures for carrying out genetic studies. See also p. 00.

Recent developments in mammalian genetics were given by: W.F. Dietrick et al. A comprehensive genetic map of the mouse genome. *Nature* 1996; 380: 149–152.

C. Dib et al. A comprehensive genetic map of the human genome based on 5,264 microsatellites. *Nature*, 1996; 380: 152–154.

Other databases

There is a Malaria Genome Project, with groups in Oxford UK and Melbourne Australia. The Malaria sequence database of proteins and genes, together with probe locations and sequence-tagged sites (STS) and restriction maps, is at the Walter and Eliza Hall institute of Medical Research in Melbourne. The WHO/TDR has launched international parasite genome networks for *Filaria*, *Schistosoma*, *Leishmania*, *Trypanosoma brucei* and *T. cruzi*.

The Nematode database of *Caenorhabditis elegans* is at the MRC Laboratory, Cambridge, UK. Known as ACEDB it contains physical and genetic map data of all the genes and cosmids, all sequences etc. *Caenorhabditis elegans. Modern Biological Analysis of an Organism*, edited by H.F. Epstein and D.C. Shakes (Academic Press, 1995) is the first laboratory handbook for this important organism. About 80% of the genes will be required by the end of 1996.

Although there is a *Drosophila* Information Database at Indiana University, the information in it and other databases has been combined in FlyBase at Harvard University. It contains all the information from the 'Redbook' – *The Genome of* Drosophila melanogaster, by D.L. Lindsley and G.G. Zimmand (Academic Press, 1992) – and more recent information. Extracts of the data are published in the research newsletter Drosophila *Information Service (DIS)* (see p. 27). FlyBase is comprehensive and contains a directory of researchers, a bibliography of more than 65 000 records, stock lists, a genetic map, chromosome aberrations, clones, cosmids and associated databases. Each Item is given a unique number. (See p. 124.)

Among the bacteria there are EcoSeq, EcoMap and EcoGene databases for *Escherichia coli* at the National Center for Biotechnology Information (see p. 140). ECD is a database for *E. coli* at the EMBL, and CGSC at Yale University contains descriptions of the thousands of mutants based on the Genetics Stock Center. Other details are given in *Microbiology Reviews* 1993: 57 part 3. Brief descriptions of physical locations and loci are printed quarterly in *Journal of Bacteriology* (see also p. 146). The *Bacillus subtilis* genome sequence project was described in *Microbiology* 1995; 141: 249–350. The genetic map of *Salmonella typhimurium* edition VIII was given in *Microbiology Reviews* 1995; 59: 241–303.

The first complete sequence of a bacterium in 1995 (see p. 128) has been followed by news that GTC of Massachusetts and Glaxo-Wellcome have sequenced 1.8 million bases of the medically important bacterium *Helicobacter pylori*. It is also reported that about 60% of *Mycobacterium leprae* and *tuberculosis* have been sequenced, but only half will be available to the public (*New Scientist* 30 September 1995, p 4). In early 1996 the complete sequencing of the 2.8 million bases of *Staphylococcus aureus* was announced. The findings are to be published in 1998, but significant sequences have been filed for European patents.

Genomic data for various plants are maintained at the National Agricultural Library (see p. 100)., The NCBI maintains a depository of databases which include the following: EcoSeq and EcoMap (see above, *E. coli*), a Restriction Enzyme Database, a Transcription Factors Database and the Eukaryotic Promotor Database (from the Swiss Institute for Experimental Cancer Research at Lausanne). There is a Proteolysis databank – defined sites for proteolytic enzymes – at the University of Bari, Italy.

Appendix

The different databases interact as shown in Figure 16.1. No descriptions have been published for:

Maize genome, MaizeDB, at the NAL;
WormPep at the Sanger Centre;
DictyDb at the University of California, San Diego;
SacchDB at the Genetics Department School of Medicine at Stanford.
The following were described in *Nucleic Acids Research* 1994; 22 Part 17:
Swiss-Prot, EPD, FlyBase, HSSP, EMBL databases, ECD, REBASE, Prosite, LISTA, OMIM and ENZYME. Other databases have also been described:
SubtiList, in *Microbiology* 1996; October and November issues
EcoGene and StyGene in *ASM News* 1993; 59: 335–341
GCRDb in *Receptor Channels* 1994; 2: 1–7
ECO2DBASE in *Electrophoresis* 1992; 13: 1014–1054
YEPD in *Electrophoresis* 1995; 16: 1170–1171
SWISS-2DPAGE in *Electrophoresis* 1995; 16: 1131–1151
TRANSFC in *Journal of Biotechnology* 1994; 35: 273–280
and PDB in *Computational Molecular Biology. Sources and Methods for Sequence* Analysis, edited by A.M. Lesk (OUP, 1988) (see pp. 69–81).

I am most grateful to Dr Graham Cameron of the European Bioinformatics Institute of EMBL for this information and the diagram.

For other databases see pp. 91 and 244.

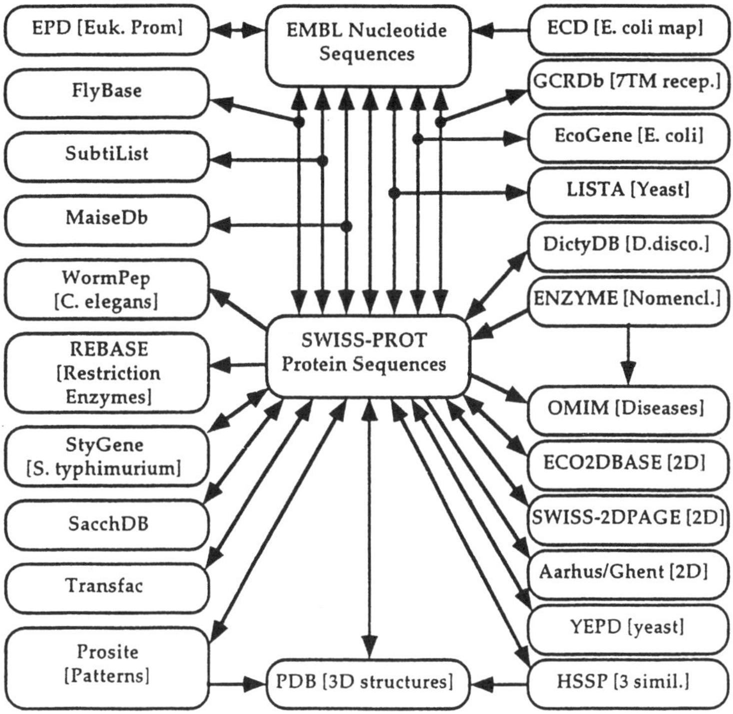

FIG. 16.1: How the databases interact.

More maps, more sequences

The progress in mapping and sequencing is astounding. A new human gene map can be found on http://www.ncbi.nlm.nih.gov/SCIENCE96/ with nearly one fifth of the estimated 100 000 genes already mapped. The UniGene collection of 50 000 clusters of sequences, each representing the transcription product of a human gene, is part of Gene-Bank of NCBI (see p. 140). A gene map of the human MHC will be published in *Immunology Today* early in 1997.

The Human/Mouse Homology Maps with 1400 loci are also available through NCBI. Three dimensional reconstructions of the mouse embryo, similar to those of *Drosophila* (see p. 126), show the action of genes active in development, using the GenexDB database at http://genex.hgu.mrc.ac.uk/

Among the several organisms completely sequenced have been the yeast *Saccharomyces cerevisiae* with 12 million base pairs and *Helicobacter pylori.* The complete sequence of *Bacillus subtilis* is expected in July 1997 (see p. 145). The techniques used at the Sanger Centre (see p. 143) for sequencing the yeast genome have been used for the reference strain of *Mycobacterium tuberculosis* whose complete sequence should be known early in 1997. It is hoped that perhaps 10 000 new targets for antibiotics and other agents will be revealed.

Chagas' disease affects more than 18 million people in Latin America and is the target of an eradication programme (see p. xv). The *Trypanosoma cruzi* genome is being mapped using cosmid libraries (Hanke et al, 1996) and it will be interesting to see if eradication is completed before the usefulness of the sequencing is apparent.

Reference

Hanke J, Sánchez DO, Henriksson J, Åslund L, Pettersson U, Frasch ACC, Hoheisel JD. Mapping the *Trypanosoma cruzi* genome: analyses of representitive cosmid libraries. *BioTechniques* 1996; 21: 686–693.

CHAPTER SEVENTEEN

Culture collections, animals and safety

H.V. WYATT

Culture collections

Culture collections have grown in importance with the growth of biotechnology and the deposit of microorganisms for patent purposes under the Budapest Treaty (see p. 137). A short account of patent protection through the deposit of microorganisms is given in *Microbiology Europe* 1995; 3 (6) Nov/Dec, 18–23, although the list of International Depositary Authorities (IDA) has no addresses.

The major culture collections are linked through the World Federation for Culture Collections (WFCC) and regionally through the Microbial Information Network for Europe (MINE) and the Information Centre for European Culture Collections (IC), for example. MINE is a database of strain cultures of bacteria, yeasts and fungi from 12 European countries with 33 institutions: data on each strain are extensive. There are also national organizations, such as the UK Federation for Culture Collections (UKFCC) and the German Collection of Microorganisms and Cell Cultures (DSMZ from January 1996). Many collections may be accessed via printed catalogues, online, an disk and through the Internet. Services offered extend to manuals, e.g. preservation methods: *Freezing and Freezedrying* (ATCC): quality control methods for cell lines (ATCC): services, e.g. elimination of mycoplasma from cell lines (DSMZ): and databases, e.g. List of Valid Bacterial Names (DSMZ). Several of these bodies produce newsletters and those of UKFCC, WFCC and the European Centre for Animal Cell Cultures may be accessed via the Internet (see p. 18), which is also linked to the Microbial Germplasm Network at the University of Oregon, USA. Many of the catalogues, including those of commercial suppliers, contain information and references to the cell lines. A recent review is a collection of papers from a conference held in August 1993, *Microbial Diversity and Ecosystem Function*, by D. Allsopp, D.L. Hawksworth and R.R. Colwell (CABI, 1995). Another international conference was held August 1996.

The Microbial Strain Data Network (MSDN) offers access to the databases and catalogues, meetings etc. through the public data networks and the Internet. Queries are answered by mail or fax. Databases include biotechnology, microbial data, cell lines and hybridomas.

The American Type Culture Collection (ATCC) is the largest and most important collection and produces many catalogues in print and diskette formats, on CD-ROM and online via the Internet, e.g. *Animal Viruses and Antisera*, *Chlamydia* and *Rickettsiae* (6th edn., 1990) which lists more than 2000 strains and antisera. Information includes a virus clone index, source of isolation, references, host range and choice, and a monoclonal antibody index of hybridomas producing antiviral antibodies. *Microbes and Cells at Work* (2nd edn., 1991) is available in print and as disk and lists over 50 000 strains, with special applications and over 5000 verified literature citations. The latest is the *ATCC Yeasts Reference Guide*, 1995 with 254 pages and descriptions of 4714 strains of 606 named species and varieties. There is a cross-index to other laboratory strain descriptions.

Many nations have national culture collections, of which the more important accept deposited cultures. The Belgian Coordinated Collections of Microorganisms (BCCM) has more than 40 000 strains and 1000 plasmids listed in four printed catalogues 1992. They supply an excellent manual for depositing biological material for patent purposes.

There are ten culture collections in the UK, linked through the UKFCC, which has a set of catalogues on disk for computer searching. A government review published in late 1996 suggested many changes, with the collections to be grouped in three major nodes. Two collections are: *IMI Culture Collection Catalogue*, (10th edn., edited and published by the International Mycological Institute, 1992), and the, National Collection of Yeast Cultures, Institute of Food Research, Norwich NR4, 7UA, UK. There are more than 2500 strains listed, with extensive information, in the *NYCC Catalogue* 1993, which is available in print or on floppy disk for Macintosh or IBM PCs.

A different collection is held by the (UK) National Institute of Biological Standards and Controls, which is a repository of research reagents for AIDS, including cell lines, viruses, recombinant viral proteins, antibodies etc. There is an AIDS Repository at the NIAID at Bethesda, Maryland, with branches in Paris and London; These hold reagents, virus stocks, cell lines and clones. A collection of more than 400 bacteria from the Antarctic has been distributed among Polish workers.

The European Society of Animal Cell Technology (ESACT) holds meetings and publishes the proceedings, e.g. the 12th meeting, *Animal Cell Technology: Products of Today, Prospects of Tomorrow*, edited by R.E. Spier, J.B. Griffiths and W. Berthold (Butterworth–Heinemann, 1994). As books date so rapidly this series should be valuable. The European Collection of Animal Cell Cultures is held at the Public Health Laboratory Service/Centre for Applied Microbiology and Research (PHLS/CAMR), Porton, UK.

The National Collection of Type Cultures UK (NCTC) sells a Quality Control Reference set of 44 cultures for laboratories in food, dairy, water, spoilage and environmental microbiology: also sold in seven smaller units. For culture collections in India see p. 103.

The Living Resources for Biotechnology series by CUP comprises: *Animal Cells*, edited by A. Doyle, B.E. Kirsop and R. Hay (1991); *Bacteria*, edited by L.R. Hill and B.E. Kirsop (1991); *Yeasts*, edited by B.E. Kirsop and C.P. Kurtzman (1988); *Filamentous Fungi*, edited by D.L. Hawksworth and B.E. Kirsop (1988); and, *Maintenance of Microorganisms and Cultured Cells*, by B.E. Kirsop and A. Doyle (2nd edn., 1991).

A specialized but increasingly important subject is covered in *Arthropod Cell Culture Systems*, edited by K. Maramorosch and A.H. McIntosh (CRC, 1994). *Cell and Tissue Culture: Laboratory Procedures*, edited by A. Doyle, J.B.G. Griffiths and D.G. Newell (Wiley, 1993) is a looseleaf manual of 900 pages with regular updates (three with a total of 100 pages were planned for the first year). This book is one of a new series by Wiley-Liss. There is a comprehensive chapter on the verification of cell identity, an extremely important subject as many cell lines have been contaminated and overgrown by HeLa cells in particular (see also below). Services for cell line verification are given in *Linscotts Directory* p. 44. The new technique of pyrolysis may used to confirm the identity of a cell line and will also identify cultures which are contaminated with mycoplasmas. It is particularly important that biochemical studies etc. using cell cultures (see p. 167) are validated by checks on the cultures before and after the experiments.

The cost of 50 pyrolysis tests is very modest. Pure strains of animals must be tested to confirm what they are said to be (but not by pyrolysis)

In vitro Cultivation of Animal Cells, by D. Gor and E. Lucassen (Butterworth–Heinemann, 1995) is an open learning text with sections on oncogenes and cancer, *in vitro* fertilization and genetic engineering, and transfection vectors. Other texts in the BIOTOL series cover plant cell culture. *Plant Cell and Tissue Culture*, edited by A. Stafford and G. Warren, is one of the Open UP Biotechnology series 1991. *Tissue Culture Techniques*, by B. Martin (Birkhauser, 1994) has safety and sterility as underlying themes, with problems and exercises. *The Preservation and Maintenance of Living Fungi*, by D. Smith and A.H.S. Onions (2nd edn., CABI, 1994) is the standard text. *The International Mycological Directory*, by G.S. Hall and D.W. Minter (3rd edn., CABI, 1994) contains information about libraries, databases, living and dried collections, visual material and directories. A number of videos on cell culture, as well as many books are sold by PGC Scientific, Box 15 Bristol BS99 5NN, UK.

HIBIO CD-Strains is a CD-ROM microorganism database which is updated yearly (Hitachi Software Engineering America, Brisbane California). The culture collections are ATCC (see above), the *List of Cultures* of the Institute of Fermentation, Osaka, the *List of Strains* of the Japan Collection of Microorganisms, and the European Collection, with others to be added. Information may be found by name, culture collection designation, GC content, growth temperature etc.

The BIOSIS register of Bacterial Nomenclature (BRBN) contains comprehensive data on nomenclature, synonyms, references etc. on 16 000 bacterial names. Tapes may be leased and the information transferred. The Taxonomic Reference File database (TRF), also from BIOSIS, is online and free (see p. 77).

The various free product news magazines contain not only useful information about products, catalogues etc., but also good articles, such as 'Incubator design and the prevention of cell culture contamination: a practical guide' in *Laboratory Products Technology* of August 1994 p. 12–13. Bibby Sterilin has a booklet on managing cell culture contamination.

A more comprehensive guide is *Cell Biology. A Laboratory Handbook*, edited by J.E. Cells (3 vols, Academic Press, 1995) which includes hundreds of tables and lists of media, cell lines, reagents and suppliers.

Safety

Accidents will happen. People become careless. The list of researchers, veterinarians and laboratory workers who have acquired brucellosis is very long: about 2% of reported brucellosis infections occur in laboratory workers. There were two small outbreaks of smallpox in England in 1973 and 1978 following the escape of the virus from well respected laboratories. *Laboratory-Acquired Infections*, by C.H. Collins (3rd edn., Butterworth–Heinemann, 1993) is an excellent book. *Laboratory Safety. Principles and Practice*, by D.O. Fleming *et al.* (2nd edn., ASM, 1994) includes the latest CDC/NIH *Biosafety in Microbiological and Biomedical Laboratories.* The simplest precautions are often flouted: pens and fingers being sucked, cultures of pathogens in the same refrigerator as the milk for the tea.

The use and abuse of animals

Many countries now have national codes of practice to protect animals, and journals often require an undertaking that experiments and studies have been made in accordance with good practice. Anyone thinking of using animals for experiments should approach the Animal Ethics Committee, the Animal Use and Care Committee or the local Home Office Inspector in the UK. If there is no such person or committee, there will still be external controls laid down by, for example, the National Institutes of Health (NIH).

The use of animals has decreased and is increasingly being replaced by bacteria, tissue and organ cultures and computer modelling. *Alternatives to Laboratory Animals*, better known simply as *ATLA*, has papers, news and views, reports of workshops and conferences, original papers and reviews of books. There is a bibliography of about 600 items per year selected from MEDLINE, mainly in toxicology. The journal is produced by the Fund for Replacement of Animals in Medical Research (FRAME) in Nottingham, UK. It is now 35 years since, at the instigation of the Universities Federation for Animal Welfare (UFAW), the 3Rs were proposed: replacement, refinement and reduction. The European Centre for the Validation of Alternative Methods is at Ispra near Milan.

The UFAW Handbook on the Care and Management of Laboratory Animals, edited by T. Poole (6th edn., Longman, 1987) appears in a new edition about every ten years. It is the premier reference book and should be in every animal house: it has references to the other more detailed books. The chapter on legislation has references to US, European and Australian requirements. *Handbook for the Animal Licence Holder* edited by J. Bunyan (2nd edn., Institute of Biology, 1991) is a working guide for use in the UK following the 1986 Animals (Scientific Procedures) Act. A short review of animal experimentation in the UK, USA, Canada, Scandinavia, France, Germany and Japan was made in the *British Medical Journal* 1993; 306: 1019–1023. A highly recommended code of practice is the *Australian Code of Practice for the Care of Animals for Experimental Purposes*, 1990. The Institute for Laboratory Animal Resources, ILAR, plans to produce a CD-ROM of information about US legislation and guidelines, and this may include material from other countries. *Alternatives to Animal Use in Research, Testing and Education*, Office of Technology Assessment, Congress of the US (Dekker, 1988) is a good reference book with US regulations and sections on selected foreign countries. A useful book is *Laboratory Animals. An Introduction for Experimenters*, by A.A. Tuffery (2nd edn., Wiley, 1995).

CRC Press publish large, expensive, but good reference reviews, such as *The Experimental Animal in Biomedical Research. Vol 1: A Survey of Scientific and Ethical Issues for Investigators*, by B.E. Rollin and M.L. Kesel (1990), which gives an American view of ethics, pain and animal care.

The Federation of European Laboratory Animal Science Associations, FELASA, holds meetings every three years, with published symposium proceedings. It also publishes recommendations on animal health monitoring for both breeding colonies and experimental animals, and on education and training. A FELASA report on pain and distress in laboratory rodents and lagomorphs was published in *Laboratory Animals* 1994; 28: 97–112. An American equivalent is *Recognition and Alleviation of Pain and Distress in Laboratory Animals*, from the Committee for . . . (National Academy Press, Washington DC, 1992). *Animal Welfare* is a quarterly journal from UFAW. *Animal Welfare and the Law*, by D.E. Blackman, P.N. Humphreys and P. Todd (CUP, 1990) is not applicable in the USA. See also p. 234–235.

It is not always recognised by young workers that the handling of animals involves exposure to animal allergens, and may sometimes lead to allergies which may be distressing and even incapacitating. Older workers should ensure that younger staff have minimal exposure to possible allergens. Although the commonest allergy is probably to rats, and allergy to guinea-pigs is a recognized hazard for immunologists, it should not be forgotten that allergies to insects are also common. *What You Should Know About Allergy to Laboratory Animals* (HMSO, 1989) is a good guide.

Laboratory Animals 'particularly encourages papers describing work' leading to the 3Rs and to improvements in the welfare or wellbeing of the animals used. It has an annual buyers' guide and publishes a number of books, e.g. *International Index of Laboratory Animals*, edited by M.F.W. Festing (6th edn., 1992) with sources of animals used throughout the world. *Linscott's Directory* lists suppliers of invertebrates, including protozoa, lower vertebrates and animals such as mice, whether inbred, F_1 hybrids, non-inbred, congenic, mutants etc., as well as animal-related custom services such as pathogen screening (see p. 44). Animals should be as free of infection as possible and, preferably, should be graded as specific pathogen free.

Animal Technology has the contents pages of journals from Europe, including Latvia and France, Israel, the USA, Cuba and India. *The Bulletin of the Institute of Animal Technology (IAT)* has a pullout suppliers' register in each issue.

Academic Press publishes many books, including those from the American College of Laboratory and Animal Science series, e.g. *The Laboratory Hamster*, edited by G.L. Van Hoosier and C.W. McPherson (1987). *Mouse News Letter*, which started as a research newsletter with a limited circulation, is now a journal published by OUP. The first journal issues contained listings of DNA probes, libraries etc., gene symbols, maps etc. There was an issue on inbred strains and Mouse Alloantigen News. *Genetic Variants and Strains of the Laboratory Mouse*, edited by M.F. Lyon, A.G. Searle, S. Raston and S.D.M. Brown 2 vols (3rd edn., OUP. 1996) is a monumental reference book which may be superseded by the database at the Jackson Laboratory (see p. 144).

Focus on Veterinary Science and Medicine is a monthly bibliography with abstracts on diskette. About 2500 items per issue come from journals, books and proceedings.

Attacks by extreme animal rights groups have been increasing and attempts have been made to counter their propaganda. In the UK the Research for Health Charities Group, Shepton Mallet BA4 4YZ, representing 11 large medical charities, produces materials for schools, speakers etc. The RDS, originally the Research Defence Society, promotes understanding of animal research in medicine, and publishes a *Directory of Animal Research Training Courses* (2nd edn., 1994) listing 38 courses. All new applicants for Home Office Personal Licences must have successfully completed an accredited training course. The European Biomedical Research Association, EBRA, was formed in 1994. In the USA the Foundation for Biomedical Research, FBR, includes among its Board members the former Surgeon General Dr C. Everett Koop, well known for his stand on HIV and AIDS.

The FBR has a newsletter and has produced a video and an *Animal Rights Directory.*

Man and Mouse. Animals in Medical Research, by W. Paton (2nd edn., OUP, 1993) is a popular, authoritative and very readable book.

CHAPTER EIGHTEEN

Biochemical and molecular sciences

A.D. SMITH

Introduction

The success of biochemistry can be measured by the efforts that have been made by other disciplines to prevent it from becoming all-embracing, but setting aside political considerations, the intrusion of a biochemical approach into every branch of biology has profound implications for a survey of what has hitherto been regarded as biochemical literature. Most significant of the stratagems that have been employed in the battle for territory has been the use of the term 'molecular' as a substitute for 'biochemical'. Indeed, if the name 'molecular biology' had not already been misappropriated for a much narrower, and entirely inappropriate, subject area, it could very appropriately have been used to designate all that area of knowledge and technique associated with the investigation of biological mechanisms at the molecular level. The fact is that the biochemical literature is now dispersed over a wide range of titles, many of which at first glance give no indication of any biochemical content. As a result, although biochemical data may still be found in publications that retain a reference to biochemistry in the title, just as often they are embedded in others that use the soubriquet 'molecular' to qualify another topic, e.g. 'molecular pharmacology', 'molecular membrane biology' – a recently announced replacement for 'membrane biochemistry'; or, in those that identify a specific topic in the title, e.g. 'Kidney', and contain biochemical contributions immersed in other papers embracing a wide spectrum of techniques. Most commonly there is an increasing tendency to identify a particular specialized area in the title, examples being 'protein science' or 'receptors and channels'. It is thus now impossible to identify concisely which journals should be classified as providing biochemical information.

A further development that is now exerting a profound influence on the dissemination of biochemical knowledge and ideas is the advent of electronic technology. Initially, this was introduced to manipulate large databases that hold and make available the data produced by indexing systems such as *Index Medicus* (MEDLINE) and research teams involved in sequencing nucleic acids and proteins. Electronic methods possess the advantage that the database can be modified rapidly and continually as new data are provided, and speed of access ensures that those interested obtain very up-to-date information. Furthermore, computer databases can hold quantities of data that on paper yield massively large and unwieldy products. Developments in international computer networks are making these databases much more widely accessible to the individual, but the information is held for the most part in flat files, having been inserted without much predigestion, so that for all except the specialist, information retrieval is a daunting task. There is thus a need for sources that provide information in a digestible, more readily available form. Thus, the electronic revolution is impinging more widely on scientific publication, a number of journals now being made available in electronic form, either online or on CD-ROM. The publisher is now increasingly an information provider. Electronic tools are valuable in meeting this need, not only in providing online information sources, but also in facilitating the organization of information in the production of books. There is a tendency now for the two technologies, electronic and paper, to merge, as is illustrated by the appearance of a number of serial and other printed publications that will be maintained in an updated state using computer technology.

I am conscious that a sister publication contains an extensive and recent survey of the literature (Rowland, J.F.B. Biochemistry, biophysics and molecular biology. In *Information Sources in the Medical Sciences* edited by L.T. Morton and S. Godbolt 4th edn., Bowker-Saur, 1992, pp. 155–190). I shall not recapitulate this information but rather comment on its use, survey trends and extend the information to the present date.

Books and serials

Textbooks stand out among the newly regenerated printed matter in that they continue to declare themselves to be biochemical.

Among the major textbooks, new editions of several were published in 1995. These include *Biochemistry*, by D. Voet and J.G. Voet (2nd edn., Wiley) *Biochemistry* by L. Stryer (4th edn., Freeman) and *Principles of biochemistry*, by G.L. Zubay, W.W. Parson and D.E. Vance, which comes with a CD-ROM (from Brown). *Principles of biochemistry*, A.L. Lehninger, D.L. Nelson and M.M. Cox (Worth) was published in 1993, as also was an edition of the same title by H.R. Horton, L.A. Moran, R.S. Ochs, J.D. Rawn and K.G. Scrimgeour, (Prentice-Hall). *Textbook of biochemistry, with Clinical Correlations*, edited by T.M. Devlin (Wiley-Liss) reached its third edition in 1992. Books of this type, along with dictionaries (see p. 37), are points of first resource in seeking biochemical information. In addition, as regards major works, the third edition of *The Molecular biology of the cell*, by B. Alberts, D. Bray, J. Lewis, M. Raff, K. Roberts and J.D. Watson, was published in 1994 by Garland. *Genes V*, by B. Lewin, was another welcome new edition from OUP in 1994. Among smaller books recently revised there are *Biochemistry*, by M.K. Campbell, with illustrations by Irving Geis, (2nd edn., Harcourt Brace; 1995), *Biochemistry Illustrated*, by P.N. Campbell and A.D. Smith (Churchill Livingstone, 1994) and *Harper's Biochemistry*, by R.K. Murray, D.K. Granner, P.A. Mayes and V.W. Rodwell (23rd edn., Prentice Hall, 1993), this latter maintaining a place despite quite primitive illustrations.

Textbooks have traditionally been backed up by more detailed monographs. A problem for books such as these is that their useful life is extremely short, and so it is hardly realistic to list works of this kind. Perhaps the most interesting development by book publishers has therefore been the production of serial books that will be regularly updated at short intervals. In the endeavour to aid scientists in grappling with the outpouring of data, especially on proteins, a number of publishers have initiated serial publications that assemble essential protein data in readily accessible form. These include the *Enzyme handbook* (Springer Verlag), which aims to provide limited essential information for all enzymes for which good data are available. It is being published in volumes each covering about 200 enzymes, grouped according to EC Number, giving systematic and common names, reaction catalysed, substrates, inhibitors, cofactors and kinetic data. Critical to the value of such a work will be the frequency with which it is updated. Another series, *Macromolecular structures* (Current Biology) details X-ray, NMR, fibre and electron crystallography structures in annual volumes. *Human protein data* (VCH) is another collection of protein data. *Protein profile*, advertised as a journal of 10 issues a year (Academic Press), is in effect a multivolume reference work, each issue containing details of a particular family of proteins, including three-dimensional structure graphics, sequence alignments and details of domains and motifs. It is claimed that each topic will be updated annually. At the rate of publication envisaged, this series will embrace only a small proportion of all protein families by the turn of the millennium (see also p. 16).

As a contribution to a recent development towards looseleaf binding, Kluwer in 1990, brought out a *Plant molecular biology manual* in looseleaf form, which has been updated by supplements. Wiley has recently engaged in a major initiative publishing looseleaf manuals of laboratory methods in the *Current Protocols* series, e.g. *Current protocols in protein science* and *Current protocols in molecular biology*, which are available on CD-ROM, to be updated quarterly. The CD-ROM provides the facility to cut and paste into wordprocessing packages, and for incorporation of the experimenter's own notes.

Several publishers have produced series which provide essential data for the research worker. These include: the *Intelligence Unit* series from Springer Verlag, e.g. *Molecular biology intelligence unit*, *Neuroscience intelligence unit*; the *FactsBook* series from Academic Press, e.g. *The oncogene factsBook*, *The extracellular matrix factsBook*; the *Essential Data* series from Wiley, e.g. *Vectors and Nucleic acid hybridization*; the *Basics* series from OUP, e.g. *DNA Sequencing* and the *Guidebook* series, e.g. *Homeobox Genes.* All have a common aim of providing essential and topical information in digestible form, at the high level of specialization characteristic of today's research. The need for such works to be maintained in a state of currency is obvious, and only those series which are updated every two or three years will be successful.

All these works suffer the same limitations: the large amount of work entailed in their production, which must be undertaken by scientists who are fully occupied on other tasks, the relatively limited funds for the purchase of such works, and the tidal waves of data being produced. These considerations point to the electronic database as the agency most able to cope with these problems, as data and abstracts prepared by experimental groups can readily be incorporated into such databases. However, the stimulus for the production of printed works of this kind is (to quote from an advertisement for *Protein Profile*) that 'the bibliographic and sequencing material within these (13 different) databases is presented separately and inconsistently, making it difficult to compare and contrast related proteins', and one may add, making it difficult if not almost impossible, for any but the scientist specializing in these areas to extract any information from them at all. There is thus a continuing need for information that has been abstracted and formalized, the only question being whether, in the face of the difficulties, it is practicable and commercially viable to produce it in this form. The actual method of production of such digested information is likely to remain the book, backed up by CD-ROM, but one can see in the regular updating of series of this type the embryonic information service of the future, as they become linked in to online systems.

An interesting venture from Elsevier, in the field of computer education, is a series of *Open Programmes for Associative Learning (OPAL)*, e.g. on *Neural communication* and *Calcium as a messenger*. These take the user through a structured programme, and could be of use to young research workers.

To satisfy the demand for accessible sources that aid the scientist in understanding new knowledge, an increasing effort is being put into the publication of dictionaries. This will become an increasingly popular activity in the coming years, as an adjunct to the reference publications in Chapter 7.

Trends in journals

New journals

Some of the titles that appeared between 1990 and 1996 and still survive are listed in Table 18.1. There has been a burgeoning of growth in two areas – molecular genetics and nucleic acid research, and neuroscience – to a lesser extent in areas such as protein science, signal transduction and cell–cell interaction, and beyond this the introduction of a number of more specialized journals.

MOLECULAR GENETICS AND NUCLEIC ACID RESEARCH

Nature Genetics appeared in a recent renewal of *Nature*'s desire to generate offspring. As might be expected, it sets a high standard and also publishes papers on genetic disease. *Human Molecular Genetics* and *Human Mutation* are both orientated towards the molecular genetics of human disease, having distinguished international editorial boards and a good quality of papers. *Transgenic Research* is dedicated to the rapid publication of research in transgenic higher organisms. *RNA* concentrates on all research connected with RNA, *DNA Sequence* is exclusively devoted to DNA sequencing and its techniques, and *PCR Methods and Applications* is another technique-orientated journal. Any doubts about the future of antisense techniques appear to have been subdued and *Antisense Research and Development* would seem to have a

Table 18.1. Journals in biochemistry and molecular biology which have been published since 1991.

Year of first issue issue	Title	Issues per year			
		1	4	6–8	10–12
1991	*Amino Acids*			+	
	Antisense Research and Development		+		
	Cellular Physiology and Biochemistry			+	
	Current Opinion in Structural Biology			+	
	Glycobiology			+	
	Journal of Biomolecular NMR			+	
	Journal of Chemical and Biochemical Kinetics		+		
1992	*Advances in Developmental Biochemistry*	+			
	Biological Signals			+	
	Cell Transplantation			+	
	Protein Science				+
	Structure				+
1993	*Acta Crystallographica D: Biological Crystallography*			+	
	Journal of Biochemical Organization		+		
	Natural Products Letters		+		
1994	*Current Biology*				+
	Letters in Peptide Science			+	
	Protein and Peptide Letters		+		
	Protein Profile				+
	Redox Report		+		
	Toxicology and Ecotoxicology		+		
1995	*Biomaterials Science and Engineering*		+		
	Journal of Molecular Catalysis			+	
	Journal of Peptide Science			+	
	RNA				+
1996	*Cell Stress and Chaperones*		+		
	Folding and Design				+
	Genes and Function			+	

lively future. *Insect Molecular Biology*, although at first sight an entomological journal, may have wider relevance in view of the application of insect cells to the expression of mammalian genes. *Journal of Marine Biotechnology* and *Molecular Marine Biology and Biotechnology* are ostensibly competing journals, and it will be interesting to see whether this proves to be the case or whether they develop distinctive characters. Finally, in this section *The*

Plant Journal is a European attempt to emulate the highly successful American *The Plant Cell.*

NEUROBIOLOGY

This is an area that seems to excite publishers into considerable activity, at least 11 new journals having appeared in the last five or six years. These include titles such as *Dementia*, ostensibly a clinical journal but which, in keeping with the modern trend, includes papers ranging from neurochemistry to case reports, and *Neurodegeneration.* Of a more general nature are *Molecular and Cellular Neurosciences, The European Journal of Neuroscience* and *Neuroreport.* The last-named introduces a new concept, of publishing very rapidly papers that make a single point and are thus very short. Does this really have a value? Only time will tell. *Molecular and Cellular Neurosciences* is competing with existing *Cellular and Molecular Neurobiology* and *Journal of Molecular Neuroscience*, and its emergence reflects the mushrooming of this field and its highly competitive nature. *The European Journal of Neuroscience* is the proprietary journal of the European Neuroscience Association, and would appear to parallel the existing *Journal of Neuroscience* already well established. For review journals there have appeared *Current Opinion in Neurobiology, Frontiers in Neuroendocrinology, Seminars in the Neurosciences* and *Comments on Developmental Neurobiology.* Covering specific areas are *Molecular Neuropharmacology* and *Journal of Neuroendocrinology.*

PROTEIN RESEARCH

An important addition to this field is *Protein Science*, a journal reporting research in protein science over a wide range of techniques, from the biophysical to the molecular biological. It is interesting in that it offers additional data and graphics in machine-readable form, an approach that is likely to become more widely used. *Structure: Form and Function in Modern Biology*, despite its grandiose title, seems destined to be preoccupied largely with protein structure. There is certainly not likely to be any shortage of material for publication in a field which generates such large quantities of data, which should allow the editors to select good-quality material, and the evidence so far is that this is the case. *Protein Expression and Purification* is a technique-orientated journal, and again there will surely be no lack of submitted manuscripts. The field is one of intense interest, and one must hope that this will be reflected in the quality of papers. *Journal of Biomolecular NMR* is another technique-orientated journal, which will cover oligonucleotides and is likely to have a large peptide content.

SIGNAL TRANSDUCTION

New entrants here are *Receptors and Channels*, *Biological Signals* and *Receptor*. Perhaps one should also include a cell biology journal *Cell Regulation*, which seems likely to have an overlap into cell signalling. Although this is one of the largest of all research areas, it is already plentifully supplied with specialist journals and it remains to be seen what these new journals will usefully supply.

CELL–CELL INTERACTION

European Cytokine Network and *Cytokine* follow specifically the cytokine trail, whereas *Mediators of Inflammation* is less specifically limited in a similar subject area (see also p. 193). *Cell Adhesion and Communication* is concerned with the fixed communicators in cell–cell interaction. Another candidate for attention is *Epithelial Cell Biology*.

SPECIALIZED JOURNALS

Researchers increasingly focus on their own immediate research, a trend that is becoming the more necessary as research becomes more competitive and the outflow of results makes it more difficult for those working at the bench to follow more general research findings. Thus the tendency is to read research papers and long reviews focused on a narrow area, and to seek more general education in short reviews. *Glycobiology*, although at first sight seeming to be specialized, caters for those who wish to stay within this field but not to confine themselves to a particular discipline. In this regard they have the existing *Glycoconjugate Journal*, which is more chemically orientated. Even more specialized are *Neuromuscular Disorders*, *Journal of Liposome Research*, *Biometals*, *Growth Regulation*, *Lung Cellular and Molecular Physiology* and *Learning Memory*, but all of these will contain biochemical papers.

COMPUTER-BASED JOURNALS

The fully electronic library, or 'library without walls', has for long remained a dream, but is about to become a reality. The increasing use of international computer networks, stimulated by developments in software connected with the Internet, is a major factor. There has been some hesitation on the part of publishers to face the need to adjust from a situation in which libraries and individuals subscribe to a paper copy, to one in which there will be provision for the library user to select and download only those articles that he needs, but a sea change in this attitude is now occurring, and a number of publishers are now actively promoting electronic journal services.

In discussing current and future developments it is necessary to distinguish simple document delivery services from more sophisticated systems that can use the full potential of international computer networks. At the level of document delivery the user simply requests a particular article, which is delivered to his computer in a convenient format. This may be HTML, the language of the Internet or, more desirably, PDF (portable document format), a format that functions as a printer-independent protocol, enabling the production on a simple (e.g. laser) printer of pages identical to those in the original article.

Several publishers now have journals online. In general, the services involve subscription to a journal or set of journals, in return for which the subscriber can access the electronic version and obtain such other connections as are provided.

For example, in some cases links are offered to other services using standard Internet procedures, such as clicking on highlighted items, which then takes the system beyond the document delivery service. Thus *Gene Combis* (Elsevier), an adjunct of the paper journal *Gene*, will provide online links to nucleic acid and protein databases. Subscription to various journals of Chapman & Hall will provide access to the online journal within the overall Thomson online system. Subscription to the printed or CD-ROM version of *Nucleic Acids Research* (OUP) will provide access to the online version. The facility to download journal articles is available in all cases, and it seems likely that links to other databases, including bibliographic databases, will soon be routine. Thus a hotlink from a bibliographic reference at the foot of an article will put on screen the abstract of the article. It is also likely that there will be expansion into other Internet facilities, such as conferencing. *BioMedNet*, an online system of Electronic Press Ltd, will initially not only make available all journals in the *Current Opinion, Current Biology* series, providing the facility to download any article from any of the journals carried for an overall subscription of $90 per year, but will also provide 'meeting rooms', 'notice boards' and a 'shopping mall'. With regard to straightforward document delivery, it is understood that beta testing will shortly begin of such provision over a wide range of journals from the Institute of Scientific Information. A number of journals are available on CD-ROM, among the first to provide such an accessory being the *European Journal of Biochemistry* and *Journal of Bio-*

logical Chemistry. Other journals offering some degree of online service include *Biochemistry*, *Chemistry and Biology*, *Journal of Cellular Biochemistry*, *Toxicology Letters*, *Toxicology and Applied Pharmacology*, *Blood*, *Eurobiologiste*, and *Immunology Today*.

Increasingly, journals are making available archival CD-ROM discs; these could be very valuable not only for libraries but also for individual scientists, but *Infection and Immunity* prices its archival CD-ROM for the previous year at $50 in the USA, and this compares favourably with, and is more realistic for the individual scientist than, *Nature*'s £200 for the previous two years. Despite all these developments it is likely that hardcopy journals will persist, but libraries may abandon the practice of maintaining bound copies, instead holding the paper copy for perhaps two years with previous issues being held on archival CD-ROM. Such changes of practice are under discussion in most large libraries.

IMPACT FACTOR

The impact factor published in the *Science Citation Index (SCI)* has become a major force in influencing trends in science journals. This is an index that purports to measure the 'value' or 'quality' of a journal, being 'a measure of the frequency with which the "average article" in a journal has been cited in a particular year' and is 'basically a ratio between citations and recent citable items published' by the journal. *SCI* suggests that the impact factor 'tends to discount the advantage of large journals over small ones, of frequently issued journals over less frequently issued ones', in that it is not just a record of total numbers of citations. There is no doubt that there is some correlation between the quality of a journal and its impact factor, particularly because those papers most avidly sought out and read by the scientific community are likely to be those most cited. It must be used with discretion, as distorting factors will undoubtedly play a part in its value (see p. 65). Thus, of the top ten titles in the list (1994), eight are review journals. This of course stems from the fact that most papers begin by citing in the introduction a review from which the reader can obtain the background to the research. However, six of these eight review titles publish heavyweight reviews of the type that reassure the author that he has offered an exhaustive coverage of the background, while leaving the reader with the feeling that something more digestible might have been preferable. Without doubt, these six are quality publications (Table 18.2), offering authoritative reviews by well selected authors. However, their impact factor by no means correlates with the frequency with which they are read, or the influence they have on thought and ideas within

Table 18.2. Review serials in biochemistry and molecular biology listed in order of impact factor (those in the top 20). From *Journal Citation Reports* 1993

1	*Annual Reviews of Biochemistry*
3	*Annual Reviews of Cell Biology*
4	*FASEB Journal*
5	*Annual Reviews of Biophysics and Biomolecular Structure*
6	*Trends in Biochemical Sciences*
8	*Advances in Enzymology*
9	*Critical Reviews in Biochemistry and Molecular Biology*
10	*Advances in Protein Chemistry*
14	*Progress in Nucleic Acid Research and Molecular Biology*
16	*Advances in Second Messenger and Phosphoprotein Research*
17	*Advances in Microbial Physiology*
19	*Advances in Carbohydrate Chemistry and Biochemistry*

the scientific community. An example of the way in which the impact factor may influence journal editors, not necessarily to the benefit of scientists, is the rumour that a journal which at the outset published excellent short reviews of the type that are eagerly read (but not much cited), has now started to publish original papers, hoping to attract highly topical papers that will be cited and thus increase its impact factor. The extent to which this weighs on a publisher is indicated by the inclusion in a recent advertisement of a new journal that it 'has an impact factor from launch'!

These reservations having been stated, the listing of the impact factor undoubtedly does give an indication of trends in scientific publishing. One of the most notable of these springs from the comments in the introduction above, concerning the move away from a 'biochemical' connotation towards a more topic-orientated status, or towards more currently favoured topics. Thus, the ratings listed in Table 18.3 indicate the degree to which the classic biochemical journals, publishing papers over the whole spectrum of biochemical research and often including papers of a more solid (rather than fashionable) nature, are under pressure from newer journals. No-one should imagine that because of this these journals are not of great value. Much of the progress of science depends on this more solid, and perhaps more critical, type of publication, which modifies earlier findings, sometimes out of all recognition. Nevertheless, editors will do well to give serious consideration to these figures. In part they denote the development of biochemical research towards a more cell-based approach, in which the subject of investigation is an intact cell, more often than not in culture (see p. 151), and frequently using the techniques of molecular

Table 18.3. Changes in impact factor rating of general biochemical journals. From *Journal Citation Reports*, under biochemistry and molecular biology

Title	1991	1992	1993
Journal of Biological Chemistry	11	14	15
Biochemistry (US)	15	20	25
Biochemical Journal	28	30	39
FEBS Letters	30	34	44
Biochem Biophysical Research Communications	27	32	45
European Journal of Biochemistry	36	35	46

biology such as the production of mutants, transfected cells or cells from knockout animals in which genes have been deleted. Often overlooked in this is the value of work done on purified proteins and other enzymes on which much of the work is based, and which will in the past have been reported in the more classic journals, but which increasingly will now be published in these more cell-orientated journals, in the hope of catching the eye of the cell biologist. Moreover, there is a tendency for purification to be aimed at sequencing the protein, with a view to cloning the gene, so that much of this work finds its way into sequence-orientated journals. The top three journals in the impact factor league table for biochemical citations are thus *Cell*, steady over the last two years analysed at the remarkably high position of 2, *EMBO Journal* at 7, and *Molecular and Cellular Biology* at 12 (Table 18.4). If one lists the most frequently cited journals, however, a different situation pertains (the difference in the positions of *Annual Reviews of Biochemistry* at 96 and *Trends in Biochemical Science* at 226, for example). This list is likely to be more useful in denoting those journals most generally of interest (Table 18.5).

Review journals

Although this chapter is devoted primarily to recent innovations, it seems useful to discuss the more important review journals. The *Annual Reviews* series remains one of the best sources of reviews of the long and exhaustive type. The review sections of *Biochimica et Biophysica Acta* are also excellent in this category, as also are the *CRC Critical Reviews* series, and series with the general titles *Progress in*... (or *Recent Progress in*....), *Advances in*.... The demand for short reviews has led to the production of an increasing number of these. A number of excellent periodical

Table 18.4. Biomolecular journals listed in order of impact factor (top 20 only).

Cell
EMBO Journal
The Plant Cell
Molecular and Cellular Biology
Molecular Biology of the Cell
Journal of Biological Chemistry
Journal of Molecular Biology
Proteins
Molecular Pharmacology
DNA and Cell Biology
Biochemistry (US)
Molecular Microbiology
Cell Growth and Differentiation
Journal of Neurochemistry
PCR Methods and Applications
Protein Engineering
Free Radicals in Biology and Medicine
Molecular Biology in Evolution
Nucleic Acids Research
Human Molecular Genetics

series fall into this category. The *Trends in* . . . series is firmly established as a major resource (see p. 5). A recent addition is *Trends in Cell Biology. Bioessays* has a high impact factor. An interesting recent development was the introduction of the *Current Opinion in* . . . series. These short reviews aim to highlight developments over the previous 12 months, concentrating on the most important aspects and indicating a few selected papers for special attention by awarding 'stars' (see p. 6). Several new branches of the *Seminars in* . . . series have emerged and been well received; each issue includes a number of reviews centred on a particular topic, with a guest editor pulling the volume together. The annual *Essays in Biochemistry* concentrates on publishing short reviews of an easily understood nature to help undergraduates. The news and views section *of Nature* is an invaluable source of highly topical surveys that are usually published simultaneously with a relevant paper.

Most of the major journals now include short reviews in their issues, very notably *Cell, Journal of Biological Chemistry, Biochemical Journal, Journal of Cell Biology, Journal of Endocrinology* and *Journal of Experimental Medicine.* Other journals that regularly include short reviews are *Biochemical Pharmacology, Biochimie, Biological Chemistry Hoppe–Seyler, European Journal of Biochemistry, Free Radical Biology and Medicine, International Journal of Biochemistry, Journal of Bioenergetics and Biomembranes* (almost a review journal), *Journal of Membrane Biology* and *Structure.* Other journals that carry more occasional reviews are *Experimental Cell Research*, *Human Genetics*, *International Journal of Protein and Peptide Research*, *Journal of Bacteriology*, *Journal of Biochemistry*, *Journal of Cellular Biochemistry*, *Journal of Lipid Research*, *Journal of Muscle Research and Cell Motility*, *Neurochemistry International* and *Nucleic Acids Research.* Some journals carry short notes on the latest developments, e.g. *Biophysical Journal* ('new and notable'), *Cancer Genetics and Cytogenetics* (lead article), *Investigative Ophthalmology and Visual Science* ('recent developments'); such inclusions are of considerable value (compare *Nature* 'news and views').

Indexing and abstracting systems

A first step in entering the literature will often be the use of an indexing and abstracting database, either as part of a continuous updating programme or to locate the literature in a topic newly of interest. *MEDLINE* and *Chemical Abstracts* are likely to be the most generally useful databases for the biochemist. It seems reasonable to comment on the limitations of these services, since without an awareness of these hopes and expectations may be unwarrantably high. First, searches will be most successful where the object of search represents a topical and clearly identifiable subject that is the focus of a clear group of researchers, and for which a readily selectable indexing term exists. For MEDLINE, for example, it is advisable to use the MESH headings (see p. 81), unless some other indexing term is known to be of use. To depart from these, and to attempt a search for terms of a more general nature, is likely to be less satisfactory because of the vastness of the database that underlies the search, and the possibility that the term being used may be employed in ways that the searcher does not envisage. Even if the search yields some measure of success, the 'noise' level is likely to be very high.

Increasingly, these services are being made available for use offline, for example on CD-ROM. An advantage of this is that most academic libraries do not make a charge for such a service, in contrast to online searches. The Institute for Scientific Information provides a version of *Current Contents* on floppy disk, for use by individual scientists on desktop and portable computers. *Current Awareness in Biological Sciences* is similarly available. As these disks are produced weekly or fortnightly, however, their use is limited to routine searching for well identified subject areas. A disadvantage of all these offline services lies in the limitations of

Table 18.5. The top 20 most-cited journals (1993).

Rating (all scientific journals)	Journal title	Citations
1	*Proceedings of National Academy of Sciences (USA)*	234319
2	*Journal of Biological Chemistry*	231324
3	*Nature*	226827
4	*Science*	178514
6	*Cell*	113325
10	*New England Journal of Medicine*	92729
11	*Lancet*	80055
13	*Biochemistry (US)*	76131
15	*Biochimica et Biophysica Acta*	66827
16	*Brain Research*	61842
17	*Cancer Research*	60626
18	*Journal of Clinical Investigation*	58670
20	*Journal of Cell Biology*	56430
22	*Biochemical and Biophysical Research Communications*	52503
23	*Nucleic Acids Research*	52218
24	*EMBO Journal*	49572
30	*Biochemical Journal*	47145
31	*Journal of Experimental Medicine*	45771
34	*Molecular and Cellular Biology*	42863
35	*Journal of Bacteriology*	40320

the small computer in speed of access, and also in its capacity for storage, so that only comparatively recent references will be available. Depending on the software used, it may be necessary to insert a succession of disks to conduct even one search. These limitations will gradually decrease as more powerful small computers become available, although the ability of libraries to modernize their facilities is likely to be restricted by financial considerations. The advent of the more widely available online journals and the searching facilities they will provide is likely to increase the use of direct online searching. Most of these facilities now also include abstracts of the paper referred to, which eases the task of selection in deciding which references may be useful.

Biochemical information for developing countries

The problem for scientists in developing countries in obtaining access to biochemical information springs not only from the limited library and database resources available to them, but also from the fact that their interests may be directed to local problems and not necessarily represented in international literature. One initiative directed at easing these problems operates as International Network for the Availability of Scientific Publications, INASP, PO Box 2564, London W5 1ZD, UK (see p. 95). Its objectives include to 'support and strengthen existing programmes involved in the distribution, local publication, exchange and donation of books, journals and related materials' and to 'encourage and support new iniatives that will increase local publication and general access to quality scientific literature'.

More maps

The study of the molecular aspects of life has turned attention away from how the enzymes work together. Maps provide a link between study of proteins, many of which are enzymes and the way that cells and whole organisms work. *Metabolic Maps*, by D.E. Nicholson which has set a record with more then 900 000 copies already distributed, will be available once more in its 22nd edn in early 1997, from Sigma, together with his map of *Inborn Errors of Metabolism.*

CHAPTER NINETEEN

Microbiology

P.J. LARGE

Microbiology is a discipline that is difficult to define. Its subject matter: bacteria, viruses and fungi, can also be tools or materials for many scientists who would not consider themselves microbiologists. Moreover, there has been a worldwide tendency to reorganize university science into larger units, which in the late 1980s led to the disappearance of some microbiology departments and a relabelling of the people who worked in them, although their work did not necessarily change. A consequence of this is that many journals and books used by microbiologists also figure in the literature of disciplines such as biochemistry, molecular biology, ecology and genetics, and do not necessarily carry the word microbiology in their titles. Similarly, journals with microbiology in their titles may contain papers that are primarily genetic or biochemical. A crude way of determining the subject matter of microbiology might be to examine the subject categories in the contents list of a general microbiological journal. *Microbiology* (formerly *Journal of General Microbiology*) lists, (microbial) development and structure, physiology and growth, biochemistry, genetics and molecular biology, genome analysis and, systematics, as well as biotechnology, environmental microbiology, plant–microbe interactions, pathogenicity and medical microbiology, and antigens and immunity.

Table 19.1 lists the top 25 microbiology journals ranked by *Science Citation Index (SCI)* impact factor and cited journal half-life, and gives an indication of the relative importance of specifically microbiological journals. Many subject lists are headed by review journals (see below). The recent tendency of journals publishing research papers to include one or more minireviews in each issue is unlikely to change these tables much, because the review citations will be submerged in the many research papers published in the same journal. No journal in the microbiology subject category listing appeared in the top 25 journals in *SCI* 1989 ranked by number of times cited in 1989. However, in the table of rankings of all cited journals by impact factor, *Microbiological Reviews* appeared as 14th. The cited journal half-life should be a significant factor for libraries to take into consideration when they review their subscription lists and which of their holdings to bind. Some journals clearly have a long shelf-life, whereas others appear doomed to instant oblivion.

Major publishers

The publishing arm of the American Society for Microbiology, ASM Press, publishes textbooks and monographs as well as symposium proceedings, laboratory methods manuals (see below) and the family of 11 well-known ASM journals, the oldest being *Journal of Bacteriology*. Six of these journals appear in Table 19.1. The other major societies, the Society for General Microbiology (SGM), the Society for Applied Bacteriology (SAB), both British, and the Federation of European Microbiological Societies (FEMS) have active publishing programmes for journals and symposia proceedings through commercial publishing houses.

Other specialist publishers include the Public Health Laboratory Service of the UK and Cold Spring Harbor Laboratory Press, each publishing a monograph series, and the latter a very useful laboratory manual series. Important commercial publishers are Academic Press (AP) (*Methods in Microbiology* and *Benchtop Manuals* series), Bios Scientific, Blackwell Scientific (*Basic Microbiology* series), Chapman & Hall, CRC Press (*Practical Handbook* series), Elsevier (*Developments in Food Science*, *Topics in Industrial Microbiology* series), Karger (medical microbiology), Springer Verlag (*Brock/Springer Series in Contemporary Bioscience*) and John Wiley (*Modern Microbiological Methods* series, *Ecological and Applied Microbiology* series).

Secondary sources

The major abstracting and current awareness services have changed little in recent years, but are becoming more important because libraries are unable to keep pace with the escalating number and cost of primary periodicals. From 1989 to 1991 most

Table 19.1. *Science Citation Index* subject category listing for microbiology, 1989. Journals ranked by impact factor. Impact factor is defined as (number of all current citations of source items published in journal *X* during the previous two years)/(number of articles that journal *X* published in those two years). Cited half-life is defined as the number of journal publication years going back from 1989 which account for 50% of the total citations received by the cited journal in 1989. Review journals are in bold type. Journals 4, 10 and 24 began too recently to have papers cited before 1985. (Data in this table are reprinted from *Science Citation Index, Journal Citation Reports* 1989, with the permission of the Institute for Scientific Information).

Rank	Title	Impact factor	Cited half-life (years)
1	***Microbiological Reviews***	16.6	4.9
2	***Annual Review of Microbiology***	9.6	6.4
3	***FEMS Microbiology Reviews***	4.6	3.1
4	*Molecular Microbiology*	4.5	1.6
5	***Reviews of Infectious Diseases*** (now *Clinical Infectious Diseases*)	3.9	4.7
6	*Journal of Bacteriology*	3.5	6.2
7	***CRC Critical Reviews in Microbiology***	3.3	5.5
8	*Antimicrobial Agents and Chemotherapy*	3.3	4.8
9	*International Journal of Systematic Bacteriology*	2.7	6.0
10	*Yeast*	2.5	3.0
11	*Applied and Environmental Microbiology*	2.4	5.3
12	*Journal of Clinical Microbiology*	2.4	4.7
13	*Journal of Antimicrobial Chemotherapy*	2.3	3.9
14	*Archives of Microbiology*	2.1	7.7
15	***Advances in Applied Microbiology***	2.1	9.9
16	***Advances in Microbial Ecology***	1.9	5.0
17	*Journal of General Microbiology* (now *Microbiology*)	1.9	7.7
18	*Systematic and Applied Microbiology* (formerly *Zentralblatt für Bakteriologie, Mikrobiologie und Hygiene* **1 Abt Originale C**)	1.8	3.8
19	*Journal of Medical Microbiology*	1.8	5.0
20	*European Journal of Clinical Microbiology and Infectious Diseases*	1.8	3.4
21	*Enzyme and Microbial Technology*	1.4	4.1
22	*Applied Microbiology and Biotechnology*	1.4	3.2
23	*Journal of Infection*	1.4	4.1
24	*FEMS Microbiology Ecology*	1.3	3.3
25	*FEMS Microbiology Letters*	1.3	4.6

publishers published more periodical titles and increased the number of pages and issues, but fewer subscriptions meant that subscription costs increased by 24%. In other words, because of

Table 19.2. Major secondary sources for microbiology

Title	Year coverage began	No. of journals covered	Estimated no. of microbiology entries (per year)	Electronic form
Biological Abstracts (A)	1926–	7 000	40 000	OL, CD
Biological Abstracts/RRM (A)	1975–		25 000	OL, CD
Index Medicus (T)†	1960–	3 000	22 000	OL, CD
Chemical Abstracts (A)*	1907–	14 000	20 000	OL, CD
Science Citation Index (T)#	1961–	3 000	65 000	OL, CD
Current Contents:				
Life Sciences (T)		1 350	about 14%	OL, disk, CD
Agriculture, Biology and Environmental Sciences (T)		950		OL, disk, CD
Current Advances (The CABS Database)	1984–			
in Applied Microbiology & Biotechnology (T)		2 000 (both)	12 000	OL
in Genetics & Molecular Biology (T)			27 000	OL
Cambridge Microbiology Abstracts Series:				
A. *Industrial and Applied Microbiology* (A)	1965–	2 100	7 700	OL, CD
B. *Bacteriology* (A)	1966–	2 100	13 200	OL, CD
C. *Algology, Mycology and Protozoology* (A)	1972–	2 100	10 000	OL, CD
Excerpta Medica Section 4. Microbiology (50% T, 50% A)+	1948–	3 500	2 500	OL, CD
Abstracts on Hygiene and Communicable Diseases (A)	1968–	1 500 (both)	3 500	OL
Tropical Diseases Bulletin (A)	1912–		3 500	OL
Yeast (T)	1993–	>200	3 000	

* Of the 80 classified sections of *CA*, the first major division (Biochemistry, nos. 1–20) would include all those relevant, of which perhaps 30% might be termed microbiological. The figure is based on the assumption that there are equal numbers of abstracts in each of the four major divisions of *CA*.
+ Now *Microbiology: Bacteriology, Mycology, Parasitology and Virology*
OL, online; CD, CD-ROM;
A, abstracts; T, titles only; Disk: diskette
† # electronic versions now have abstracts

budget limitations libraries are buying fewer journals even though the range of titles is increasing.

Table 19.2 summarizes the major secondary sources. *Biological Abstracts* covers research papers in the primary literature, and *Biological Abstracts/RRM* covers meetings literature, reviews, books and research reports. Although many people's access will be limited to those secondary sources available locally, the relevance of *Chemical Abstracts* and *Index Medicus* should not be underestimated. Their coverage is much wider than might be implied by their titles. Often, too, the yield from *Biological Abstracts* is disappointing. Generalization is difficult because of the wide spread of interests that might be termed microbiological, and individual research workers are recommended to find out which secondary sources are most useful. The secondary publications most useful to medical bacteriologists were surveyed by Carson and Wyatt (1982), who showed that *Index Medicus*, *Microbiology Abstracts*, *Biological Abstracts* and *Current Contents Life Sciences* had the best coverage of selected papers from a wide spectrum of journals. Since that time the *Current Advances* series has come into existence, but no more recent survey data seem to be available.

Most abstracting services are available on CD-ROM (Table 19.2). Online searching of secondary sources is becoming more important as reduced library budgets make fewer such abstracting publications available locally, whether in print form or as CD-ROM. If an online search is necessary, careful advice on the choice of search terms is recommended. The publishers of the *Microbiology Abstracts* series (Table 19.2) provide subscribers to the printed serial with electronic access via the Internet to the current and five previous years of that title, at no additional cost.

The increasing number of journals which now appear on CD-ROM as well as in print seems likely to delay the advent of fully fledged electronic publishing (non-print publications, see p. 17). However, the technology to store and deliver documents and articles between institutions already exists, and such services are likely to develop rapidly in the next five years (Joint Funding Councils, 1993). Bioline Publications allows electronic readers online access using gopher software without cost or registration, to a large database of abstracts, summaries and contents lists of journals, conference proceedings and similar material, including information and newsletters of the European culture collections. Such material can be downloaded and copied (see p. 18). Another recent development is UnCover Reveal, in which users can set up a user profile of journal titles and have the appropriate tables of contents automatically sent to them by e-mail at no charge. Copies of documents of interest can then be ordered by e-mail.

Current Contents Life Sciences and *CC Agriculture, Biology and Environmental Sciences* users have to select the one most useful for their work. The diskette version of *Current Contents* contains abstracts. *CC on CD-ROM* appears weekly with an annual archival disc. In addition to journal contents tables, it includes searchable author abstracts and keywords from the bibliographies of the papers. A journal-tagging feature enables libraries to label the source journals that they subscribe to. An equivalent service can be obtained online via the networked *SCI* database, by selecting 'Issues' on the opening menu.

Current Contents is one of the fastest current awareness services, although the classified approach with good cross-indexing used by the *Current Advances* database is extremely valuable. In 1992 the biotechnological and molecular biological aspects of microbiology were split into separate *Current Advances* titles, to the great inconvenience of many users.

Some specialized journals offer broadly based current awareness data in their own field, for example *Yeast* (Table 19.2).

For users with unlimited incomes, several of the abstracting and current awareness services offer to supply the original articles identified by searchers. This is clearly an area where, subject to the copyright problems being solved, there is great potential for electronic article delivery.

Books can of course be traced via the usual bibliographic services, such as Whitaker's *Bookbank* or Bowker's *Books in Print* (see p. 31), or, for those with Internet access, the searchable databases of the US Library of Congress (URL: http://lcweb.loc.gov/homepage/lchp.html) or of the Internet Bookshop (URL: http://www.bookshop.co.uk).

Widely used books in microbiology

Large reference books are still important in microbiology. The classic *Topley and Wilson's Principles of Bacteriology, Virology and Immunity*, edited by M.T. Parker and L.H. Collier is now in its eighth edition (5 vols, Edward Arnold, 1990) and very useful for its historical aspects. 'Mackie and McCartney', now known as *Medical Microbiology, a Guide to Microbial Infections: Pathogenesis, Immunity, Laboratory Diagnosis and Control*, edited by D. Greenwood, R.C.B. Slack and J.F. Peutherer (14th edn., Churchill Livingstone, 1992) and *Clinical Microbiology*, edited by E.J. Stokes, G.L. Ridgway and M.W.D. Wren (7th edn. Edward Arnold, 1993) are both popular. The mammoth *The Prokaryotes: a Handbook on the Biology of Bacteria*, edited by A. Balows *et al.*, is now in a second edition of four volumes (Springer-Verlag, 1992). *Bergey's Manual of Determinative Bacteriology*, edited by J.G. Holt *et al.* (9th edn., Williams and Wilkins, 1994) is a one-volume laboratory version of *Bergey's Manual of Systematic Bacteriology* (4 vols, Williams and Wilkins, 1989) (see also TRF p. 77). A newer guide, with emphasis on molecular taxonomic methods, is *Handbook of New Bacterial Systematics*, edited by M. Goodfellow and A.G. O'Donnell (AP, 1993) (see also p. 142). Another important series by very authoritative authors is *The Bacteria*, edited by I.C. Gunsalus and R.Y. Stanier (AP) which began in 1960 as a five-volume series, now only of historical value. A revised series began in 1979 with volume 6, *Bacterial Diversity*. The most recent title, volume 12, *Bacterial Energetics*, appeared in 1990. The *CRC Handbook of Microbiology* (2nd edn., 1978–) continues to produce new volumes, of which the most recent is volume 9 (1988), and it has spawned a new condensed version, the *CRC Practical Handbook of Microbiology*, edited by W. O'Leary (CRC Press, 1989). Academic Press has recently brought out the *Encyclopedia of Microbiology*, edited by J. Lederberg (4 vols, 1992); it has 2100 pages in total, but its contents are in some respects disappointing, because its coverage is patchy and not comprehensive.

Reviews and bibliographies

The most influential review publications in microbiology appear in bold type in Table 19.1. Other useful review series include *Current Topics in Microbiology and Immunology*, *Progress in Industrial Microbiology*, *Clinical Microbiology Reviews*, *Advances in Biochemical Engineering/ Biotechnology*, *Developments in Food Microbiology* and *Advances in Microbial Physiology*. *Annual Reviews* will supply any article from any volume by telephone (or fax) order, for a fairly modest price. Material of interest to microbiologists appears in several titles of the series, in addition to *Annual Review of Microbiology*. The tendency for journals publishing research papers to have minireviews has spawned a series of published collections of such reviews. The minireviews in *Journal of Bacteriology* have been published by ASM Press as *Frontiers in Microbiology*.

The last few years have seen the development of a different type of current awareness series, in which small-scale reviews are published in journals designed for the less specialized reader. The pioneer series of this type is the *Trends* series. *Trends in Genetics* and *Trends in Biotechnology* have important microbiological material, often suitable for students. This successful series was supplemented in 1993 by the 12th title, *Trends in Microbiology* monthly, with a diary of meetings, which covers virulence, infection and pathogenesis. Similar in some ways to the *Trends* series are the *Current Opinions in Immunology, Biotechnology* and *Genetics and Development*. Confusingly, there is also a current awareness series called *Research Trends*, which covers environmental and food science topics (see p. 83).

Classic books in microbiology

Some important books in microbiology have remained in print or been reprinted because of their historical interest long after the novelty of the work reported has been forgotten or superseded. Such books include Cecil Clifford Dobell's *Anthony van Leeuwenhoek and his 'Little Animals'*. Van Leeuwenhoek's original letters describing his discoveries are virtually impossible to find in the libraries of non-Dutch speaking countries. Similar books by and about Louis Pasteur, Lord Lister and Robert Koch are difficult to find, and are often worth searching for second-hand.

Other classic books that have been reprinted include Marjory Stephenson's *Bacterial Metabolism* and Jacques Monod's *Recherches sur la Croissance des Cultures Bactériennes* and Van Iterson, Den Dooren de Jong and Kluyver's biography *Martinus Willem Beijerinck, His Life and Work*. These books are of great value to historians of microbiology as well as students interested in the development of the subject.

Other important books are those on particular groups or species of microorganisms, which may offer valuable historical interest even when the science in them ceases to be new. A partial list is given in Table 19.3.

Methods and manuals

Methods books may date quickly and the usefulness of older volumes depends very much on the subject involved. Classic microbiological techniques remain useful for much longer than those of molecular biology. The pioneer series *Methods in Microbiology* 1969–, (AP) reached volume 25 by 1996. A classic laboratory work which reappeared after a lapse of 20 years is *Cowan and Steel's Manual for the Identification of Medical Bacteria*, edited by G.I. Barrow and R.K.A. Feltham (3rd edn., CUP, 1993). The successful *Manual of Methods for General Bacteriology* was reissued in 1993 as *Methods for General and Molecular Bacteriology*, edited by A.L. Demain and N.A. Solomon (ASM Press). Also useful are *Manual of Clinical Microbiology*, edited by P.R. Murray *et al.* (6th edn.,) ASM Press, 1995, *Manual of Environmental Microbiology* ASM Press 1996, edited by C.J. Hurst *et al.* and *Quality-control: Principles and Practice in the Microbiology*

Table 19.3. Books on particular species or groups of microorganisms published since 1987

Author or editor	Title	Date of publication	Publisher
Rose A.H., Harrison, J.S	*The Yeasts* 2nd edn (6 vols)	1987–1995	AP
Broach J.R. *et al.*	*The Molecular and Cellular Biology of the Yeast* Saccharomyces (3 vols)	1991	Cold Spring Harbor Laboratory
Biotechnology Handbooks	series		Plenum
Peberdy J.F.	*Penicillium and Acremonium*	1987	
Harwood C.R.	*Bacillus*	1989	
Minton N.P., Clarke, D.J.	*Clostridia*	1989	
Tuite M.F., Oliver S.G.	*Saccharomyces*	1991	
Murrell J.C., Dalton, H.	*Methane and Methanol Utilizers*	1992	
Mann N.H., Carr N.G.	*Photosynthetic prokaryotes*	1992	
Smith J.E.	*Aspergillus*	1994	
Barton L.L.	*Sulfate-reducing Bacteria*	1995	
Sharp R.	Thermus *species*	1995	
Neidhardt F.C. *et al.*	Escherichia coli *and* Salmonella typhimurium: *Cellular and Molecular Biology* 2nd edn (2 vols)	1996	ASM
McFadden J.	*Molecular Biology of the Mycobacteria*	1990	AP
Galli, E *et al.*	Pseudomonas: *Molecular Biology and Biotechnology*	1992	ASM
Maniloff J.	*Mycoplasmas: Molecular Biology and Pathogenesis*	1992	ASM
Sonenshein A.L. *et al.*	Bacillus subtilis *and other Gram-Positive Bacteria: Biochemistry, Physiology and Molecular Genetics*	1993	ASM
Dworkin M, Kaiser D	*Myxobacteria II*	1993	ASM
Bryant D.A.	*The Molecular Biology of the* Cyanobacteria	1994	Kluwer
Wachsmuth I.K. *et al.*	Vibrio cholerae *and cholera*	1994	ASM

Laboratory, edited by J.S. Snell, J.D. Farrell and C. Roberts (Public Health Laboratory Service Publications, 1991). Essential on the shelves of any microbiology laboratory are the various standard reference works on culture media: the Oxoid handbooks *The Development, Manufacture and Control of Microbiogical Culture Media* and *The Oxoid Manual* (Unipath Ltd, Basingstoke, UK) and the *Difco Manual* (Difco Laboratories, Detroit, USA).

Some methods books are extremely useful but others are rather uncritical compilations. The last few years have been characterized by many series of laboratory methods manuals, such as the *Benchtop Manuals* series of AP, the *Practical Approach* series of IRL Press, the *Laboratory Books* series of Wiley (including *Current Protocols in Molecular Biology*), the *Laboratory Manuals* series of Cold Spring Harbor Laboratory Press, with strain kits for the microorganisms used, the *Introduction to Biotechniques* series of Bios Scientific, also publishers of the *LabFax* databooks series, the *Protocols* series of Humana Press (e.g. *PCR protocols*, *DNA sequencing protocols* etc.) and numerous manuals published by the ASM, and the *Technical Series* of the SAB (Blackwell Scientific). A new pocket-sized series from Wiley is called *Essential Data Series*, the first title of which (on centrifugation) appeared in 1994. The classic *Molecular Cloning: a Laboratory Manual*, by T. Maniatis, E.F. Fritsch and J. Sambrook, (Cold Spring Harbor Laboratory Press, 1982) was a bestseller. There was a second edition by J. Sambrook, E.F. Fritsch and T. Maniatis, 2 vols, 1988.

Book reviews and meeting notices/reports

Reviews of new books appear in *SGM Quarterly* and *ASM News*, and notices of forthcoming scientific meetings appear in these two periodicals as well as *Trends in Microbiology* and *FEMS Newsletter*.

References

Carson J., Wyatt H.V. Retrieving references on microbiology: which secondary publications can help? *ASM News* 1982; **48**: 5–9.

Joint Funding Councils, Report of Libraries Review Group (The Follett Report) Bristol: Higher Education Funding Council for England, 1993.

CHAPTER TWENTY

Virology and immunology

H.V. WYATT

Immunology and virology have a number of features in common, the chief of which is that a most important source of new information comes from medical and veterinary use. Enormous resources have been poured into the problem of HIV infection and AIDS, the results of which will prompt new ideas of more general application in both subjects. The successful eradication of smallpox has prompted the attempt to eradicate poliomyelitis, and the extended programme of immunization (EPI), has prompted new thinking and research on the vaccines used in it. There are many conferences and initiatives.

The two specialized secondary serials, *Immunology Abstracts* and *Virology and AIDS Abstracts* (previously *Virology Abstracts*) have similar characteristics and are processed from the same database. For general and medical aspects both subjects are well covered by *Index Medicus. Current Advances in Immunology and Infectious Diseases* is likely to cover fewer of the medical aspects. *Abstracts on Hygiene and Communicable Diseases* and *Tropical Diseases Bulletin* are more clinically inclined.

Seminars in Virology 1990– and *Seminars in Immunology* 1989– each have six issues a year, each covering a particular theme, chosen by a guest editor. The annual symposia of the SGM provide excellent reviews.

Current Topics in Microbiology and Immunology is a series which has as many as 13 volumes in a single year, or as few as two. The range of topics is very wide, e.g. Apoptosis in immunology 1995 vol 200, Measles virus 1995 vol 191, and Cytotoxic T-lymphocytes in human viral and malaria infections 1994 vol 189.

Reviews of books on both subjects are published in *ASM News.*

There are several series of books which cover both virology and immunology: The *Medical Perspectives* series (Bios) includes *Cytokines, Autoimmunity, Molecular Virology, Antibody Therapy, Antiviral Therapy* and *Human Vaccines and Vaccination*. The *Immunology and Medicine* series (Kluwer) includes, for example, *Immunology of Infection* edited by J.G.P. Sissons, L.K. Borysiewicz and J. Cohen (Vol 25, 1995).

The *Molecular Medical Sciences* series (Wiley) includes *Vaccine Design*, by F. Brown *et al.* (1993) and *Molecular Biology of Immunosuppression*, by A. Thomson (1992). There is also the *In Focus* series on immunology (IRL) and, the *Labfax* series (Bios), which has books on virology, immunochemistry, plant molecular biology and cell culture.

Virology

Viruses are important in their own right as agents of disease, and the discovery of AIDS and its causative agent HIV has stimulated an immense effort. The need for a vaccine against HIV and better vaccines against polio and other viruses has increased interest in vaccines and the immunology of viruses. Viruses are increasingly used as vectors for insertion of nucleic acids into cells, and so are an important part of biotechnology, genetics and biochemistry. Genes from animal viruses can be inserted and expressed in plant viruses, and the polypeptide products can be harvested from the infected plants and purified. This can be done with, for example, the cowpea mosaic virus, using a chimaeric viral particle. The tobacco plant may be used with a genetically altered tobacco mosaic virus (TMV) to produce thousands of doses of vaccine from each kilogram of tobacco leaf protein. Antibodies, called plantibodies, can also be produced in plants. Viruses are therefore of fundamental interest and the literature is very widely scattered. *Vaccine* is now online (see p. 138). On the Internet there is the World Wide Web server for virology.

Virology. An Information Profile, by R. Nicholas and D. Nicholas (Cassell, 1983) is now very dated: the lists of journals and other information are uncritical.

Book reviews are published in *SGM Quarterly*, but virologists are tardy as the November 1996 issue contained reviews of eight books published in 1994 and seven from 1995; the August 1995 issue reviewed a symposium held in 1992 and published in 1993. A few books are reviewed in *ASM News*. *Reviews in Medical Microbiology* has some items on viruses and *Reviews in Medical Virology* 1991– promises to discuss controversial topics. *Trends in*

Microbiology includes viruses. *International Antiviral News* 1992– has abstracts of journal articles from a very wide range of subjects, selected review titles and news of meetings.

The *Encyclopedia of Virology*, edited by R.G. Webster and A. Granoff (3 vols, Academic Press, 1994) (also on CD-ROM in 1996) will probably supersede *Virology*, by L.H. Collier and M.C. Timbury (1990) the fourth volume of the 8th edition of *Topley and Wilson's Principles of Bacteriology, Virology and Immunity* edited by M.T. Parker and L.H. Collier (5 vols, Edward Arnold, 1990). The encyclopaedia is very good on molecular biology, but has little on ecology, epidemiology, pathology etc. *Topley and Wilson* is still very good on historical aspects. Books on virology now seem to be of two kinds, those like *Fundamental Virology*, edited by B.N. Fields and D.N. Knipe (2nd edn., 2 vols, Raven Press, 1991) are very good on the complicated molecular biology and viral genetics, and the books on medical etc. virology which deal more with ecology.

A good beginning is *Introduction to Modern virology*, by N.J. Dimmock and S.B. Primrose (4th edn., Blackwell, 1994), which has an appendix of inhibitors used in virology. *The Biology of Viruses*, by B.A. Voyles (Mosby, 1993) consists of a series of questions with explanatory essays. There is an appendix of the virologist's toolkit, which includes the common features of viruses. *Virology*, by J.A. Levy, H. Fraenkel-Conrat and R.A. Owens (3rd edn., Prentice Hall, 1994) is a good general text which includes plant viruses. There is a useful postscript on the origin and evolution of virus families, and short descriptions of the vertebrate and insect virus families. Among the general books *Medical Virology*, by D.O. White and F.J. Fenner, is the fourth edition of a popular book (Academic Press, 1994). Another popular text is *Viral Infections of Humans: Epidemiology and Control*, edited by A.S. Evans (3rd edn., Plenum Press, 1989).

Plenum has a series *The Viruses* 1992– with, for example, *The Arenaviridae*, edited by M.S. Salvato (1993), *The Coronaviridae*, edited by S.G. Siddell (1995) and *The Retroviridae*, edited by J.A. Levy (4 vols 1992–1995). Volume 4 of the last-named is concerned with the human retroviruses. Volume 3 contains a comprehensive 130-page review of simian retroviruses and would complement *Simian Immunodeficiency Virus*, by N.L. Letvin and R.C. Desroiers (1994), volume 188 of *Current Topics in Microbiology and Immunology*.

The Potyviridae, by D.D. Shukla, C.W. Ward and A.A. Brunt (CABI, 1994) is a review of the group which contains 30% of all plant viruses and is the most important economically. (For a listing of the viruses which infect insects, and of the properties of phage particles, see p. 39.)

The Human Herpesviruses. Biology, Pathogenesis, and Treatment, edited by B. Roizman, R.J. Whitley and C. Lopez (Raven Press, 1993) shows the nature of virology research: clinical manifestations, molecular biology and vaccine developments.

Viruses of Plants, by A. Brunt, K. Crabtree, A. Gibbs and L. Watson (2 vols, CABI, 1995) is a database of 1000 viruses worldwide in 1500 pages. It is a product of the Virus Identification Data Exchange (VIDE) project in which over 500 characters of each virus are sought (see also p. 103). The International Committeee on Taxonomy of Viruses, ICTV, publishes a report every three years and is developing a universal database which will contain information about sequences and antiviral drugs. The sixth report, *Viral Taxonomy*, by F.A. Murphy *et al.* (Springer, 1995) has details of more than 4000 viruses with micrographs, virion diagrams, gene maps, molecular biological data and tables.

Secondary serials

The major services *Biological Abstracts* and *Index Medicus* cover the general and medical aspects of virology, respectively, and the CABI serials will carry information about plant viruses. *Virology and AIDS Abstracts* covers man, animals, invertebrates, fungi etc. and the higher plants, and includes the slow viruses. Although some abstracts appear rapidly within six months, others take a year or more. Some Russian and Indian abstracts take up to two years. Some of the long delays seem odd, as they are from well known journals which should be processed quickly.

From *Excerpta Medica* and EMBASE comes Section 4: *Microbiology: Bacteriology, Mycology, Parasitology and Virology*, which until 1992 was only microbiology. Articles appear quickly even though abstracts are given. Unfortunately, the sections do not cumulate like *Index Medicus* and there are only about 2500 abstracts a year.

Abstracts on Hygiene and Communicable Diseases and *Tropical Diseases Bulletin* are both excellent abstracting serials with comments by experts: *Public Health News* and *AIDS Newsletter* are current awareness serials (see p. 85). All have extensive cover of virological topics.

As always, recent reviews and bibliographies are good starting points, but are liable to appear in journals and serials which are not in the main stream. Thus there is a 40-page review on hepatitis B immunization, with a bibliography of 155 recent papers, in a special issue of *Les Bulletins du Centre International d'Enfance*,

Vaccinations/Immunizations/Vacunaciones 1994; 38. The review is repeated in French and Spanish. About ten hepatitis viruses have been discovered and there is intense interest in them. *Hepatitis C Virus*, edited by H.W. Reesink (Karger, 1994) is a recent book and *Viral Hepatitis Reviews* 1995– is a new quarterly which will 'encourage lively debate'. The first issue in September 1995 contained an article on the standardization and quality control of assays for hepatitis B virus DNA (see also p. 47).

Virus Research has published the complete listing of viral and phage sequence references and accession numbers from the EMBL database, with regular updates. Gene sequences from GenBank are published in *Papillomavirus Report* 1990–, which is a bibliography of abstracts from the current literature from 1000 journals. Each issue, every two months, now contains a critical review, editorial comments, letters and conference reports. A sister serial is *Epstein–Barr Virus Report* 1994–.

Methods

New methods for detecting viruses are important in the fight against disease, rapid methods being particularly important, with a European society holding frequent meetings and sometimes training sessions. Electron microscopy, immunological methods, PCR (Polymerase chain reactions), centrifugation, culture and combinations of these methods are useful and have distinctive applications. *Molecular Methods for Virus Detection*, edited by D.L. Wiedbrauk and D.H. Farkas (Academic Press, 1995) and *Virology Methods Manual*, edited by B.W.J. Mahy and H.O. Kangro (Academic Press, 1995) are recent books. *Clinical and Diagnostic Virology* journal has papers on the detection of viruses.

Virology Labfax, edited by D.R. Harper (Bios, 1993) is an extraordinary compendium which would be useful in any laboratory. The incubation period for poliomyelitis was given as 3–6 days, although the accepted times are 7–28 days and the infectivity was given as 1–2 days before, and several weeks after, onset of illness, instead of a more usual interpretation of up to four weeks before paralysis and many weeks after. As some diseases become rarer in temperate countries, more and more distortions appear in the literature and myths are repeated from book to book. The *Labfax* is, nevertheless, a wonderful store of knowledge. There are sections on sequence data, antivirals, safety, immunology, monoclonal antibodies (MAb), expression of viral genes, detection systems for viruses etc. The section on cell culture has notes on contaminants and how to deal with them. The table of cell cultures includes viral susceptibilities and suggested HeLa cell contaminations (see

p. 151). There are appendices of inhibitors and of human and veterinary vaccines. A very good buy.

The American Type Culture Collection (ATCC) has catalogues of *Animal Viruses and Antisera, Chlamidiae, and Rickettsiae* and *Plant viruses and antisera* and the *Virus Isolation and Identification Guide* for 23 pathogenic virus groups, from the Centers for Disease Control (CDC) (see p. 195). Cell lines are available from ATCC and the European Collection (see p. 149), which also lists books on cell culture and through *Linscotts Directory of Immunological and Biological Reagents* (p. 44). Tissue cultures are used for the commercial production of vaccines but are liable to bacterial and other contamination, with great financial loss. Clean robotic manipulators can be used to carry out complicated changes of media and harvesting in sterile environments, without the presence of human operators. Over 30 such units are now used for human and veterinary vaccines, hormones and gene therapy products.

HIV and other exotic viruses

The discovery of HIV, the more recent outbreaks of Ebola virus in Central Africa and among primates in the USA, and the illness and death of horses and humans in Australia due to an equine morbillivirus, have focused attention on the possible spread of very nasty viruses in new environments. *Exotic Viral Infections*, edited by J. Porterfield (Chapman & Hall, 1995) discusses some of the known viruses.

The AIDS Research Database, at the National Institute of Biological Standards at South Mimms in the UK, and the AIDS Public Information Data Set at CDC, Atlanta, are about cases with raw data. The HIV sequence database is at Los Alamos, New Mexico, which supplies a compendium *Human Retroviruses and AIDS*, edited by G. Myers *et al.* which has floppy disks with updates compiled from the database. *AIDS Bibliography*, from the NLM, has up to 1000 citations per month to journal articles, books and audiovisual materials, including foreign-language publications. Among the subject headings are reverse transcriptase, gene products, T-lymphocyte subsets etc. An analysis of early HIV information is given in Chapter 10.

Prions

Papers and letters about prions may may be found in *Nature* and in the medical weeklies because of the important medical links with Creutzfeldt–Jakob disease (CJD) and BSE and scrapie. News about these diseases is also given in *Public Health News*. A recent comprehensive review is provided by the symposium on Slow

Infections of the CNS in *Annals of the New York Academy of Science* 1992; 724. There is still much controversy about the possible link between viruses and neurological diseases such as multiple sclerosis (MS) and motor neurone disease (MND or ALS). Bibliographies of these diseases may be available from time to time from the groups working in the National Institutes of Health (NIH). *Virology and AIDS Abstracts* includes a small separate section on prions and slow viruses (see p. 80).

Immunology

In a bibliography of 1100 references from 1991 on *Tuberculosis* (CABI, 1995) more than a third deal with immunological aspects of the disease. An indication of the complexity and involvement of immunology in other subjects is given by the interaction of the malarial parasite with man. Tumour necrosis factor alpha (TNF α) is involved in cerebral malaria. Particular HLA variants offer protection against severe malarial anaemia. Sequences of malarial antigens expressed in the liver stage of infection have been synthesized to find out which bind to the HLA site: cytotoxic T cells are thought to be involved. However, these associations are thought to be regional and at least eight HLA types may be involved with at least four malarial peptides. Experiments are now being made using vaccinia virus systems, lipid-conjugated peptides, direct injections of genes or virus-like peptides based on a yeast retrotransposon to induce immunity in animals. These studies require a good knowledge of the malarial genome and its products (see p. 144). Other approaches, which are not immunological, include transfection of novel genes into mosquitoes and studies of how mosquitoes detect chemicals to find their prey.

An analysis of basic immunology research was reported in *Immunology Today* 1995; 16: 413–417, and showed that the most important centres in the UK were, in order, Oxford, central London, Cambridge and west London. The survey was based on papers published in ten journals. In UK these were *Nature*, *Immunology* and *International Immunology*, and in the US *Science*, *Cell*, *PNAS*, *Journal of Immunology*, *Journal of Experimental Medicine*, together with *European Journal of Immunology*, and *Immunogenetics*. Not surprisingly, in view of the journals chosen, 60% of the papers were from the USA with 30% from the European Union countries. The UK had about 10% of the papers, ahead of both France and Germany. The choice of journals is interesting as the two weeklies and *PNAS* do not feature in the *Journal Citation Reports* under immunology, nor are the others the only top-scoring ones in impact factors. I would have thought that immunology was one of the most difficult subjects to divide into basic and non-basic. It will be interesting to see whether these results will be similar to a study of citations to the papers, and others, in the next few years.

Antibodies can be produced by transgenic plants, yeasts, bacteria etc., and phage may be used to display antibody fragments: see *Antibody Engineering*, by CAK Borrebaeck (2nd edn., OUP, 1995). See also p. 184.

Information about immunology is required by two different groups of scientists. There will be many who want to use an immunological technique and who want to know a little more about the systems they use, and of other features such as TNF or cytokines. A dictionary will help (p. 39), but for more detailed information the *Encyclopedia of Immunology*, edited by I.M. Roitt and P.J. Delves (3 vols, Academic Press, 1992) has hundreds of very short, good, informative articles. An elementary overview is *Life, Death and the Immune System*, a *Scientic American* reprint (Freeman, 1994) of the very popular September 1993 issue; a previous similar reprint was a good account of different aspects of AIDS. *Introducing Immunology*, by N. Staines, J. Brostoff and K. James (2nd edn., Mosby, 1993) is a good introduction and has a useful glossary of terms; a more comprehensive text, *Cellular and Molecular Immunology*, by A.K. Abbas, A.H. Lichtman and J.S. Pober (2nd edn., Saunders, 1994) has no glossary but has an appendix of the principal features of known Cluster Differentiation (CD) molecules. *Immunology*, by C.A. Janeway and P. Travers (Garland/Blackwell, 1994) is very good, with an excellent index and good references. It is good on autoimmunity and the humoral immune response.

Immunology Today has excellent illustrations, most in colour, a full list of conferences, excellent articles and a few book reviews. However, it is not for the uninitiated as the articles are very dense indeed and full of abbreviations. Some of the larger and well illustrated textbooks are also very complicated for the outsider. *Immunology at a Glance*, by J.H.L. Playfair (5th edn., Blackwell, 1992) is clearly a popular book although I find it difficult: immunology is complicated. It comprises 40 diagrams, each with one to two pages of explanations, mainly in the form of short descriptions and definitions. There is also a diagram of comparative molecular weights.

Immunology – a Comparative Approach, edited by R.J. Turner (Wiley, 1994) covers the defence mechanisms of plants and the immune systems of invertebrates, fishes, amphibians, reptiles and birds, and mammals. One reviewer has suggested that this interesting book is rather detailed but will be useful for reference. *Principles of Cellular and Molecular Immunology*, by J.M. Austyn and K.J. Wood (OUP, 1994) has been recommended as a good general textbook which assumes less prior knowledge than many others. However, there is little treatment of clinically relevant topics. References extend into 1992.

Complement: in Focus, by S.K.A. Law and K.B.M. Reid (2nd edn., OUP, 1995) is a short guide to a complicated cascading system whose genetics and control have been of great interest. *Immunoglobulin Genes*, by T. Honjo and F.W. Alt (2nd edn., Academic Press, 1995) includes human, mammalian and other vertebrate systems.

Secondary serials

Immunology Abstracts has 13 000 abstracts a year. Some appear very quickly, but others even from important journals, may take up to 18 months. *Excerpta Medica* Section 26 and EMBASE *Immunology, Serology and Transplantation* is fairly rapid and has 2600 abstracts per annum. There is a subject index in each issue and each year. EMBASE CD: IMMUNOLOGY AND AIDS is a CD-ROM of a rolling 10-year collection of about 520 000 records. *Current Advances in Immunology and Infectious Diseases* has 26 000 items each year from over 2000 journals. The journal base is very wide, e.g. *Trends in Ecology and Evolution* and *Progress in Fish Research*.

Abstracts on immunological subjects will also appear in other secondary serials, e.g. *Genetics Abstracts*, *Human Genome Abstracts* and *Nucleic Abstracts*.

Methods

There are many books and journals about methods (see p. 41). *Immunoanalyse et Biologie Spécialisée* has articles in French and abstracts in French and English. It aims to follow advances and provide information about practical problems.

The *Manual of Clinical Laboratory Immunology*, edited by N.R. Rose *et al.* (4th edn., ASM, 1992) is a complete reference work. *A Practical guide to ELISA*, by D.M. Kemeny (Pergamon, 1991) is a short book for the newcomer to this basic technique, with notes on materials and troubleshooting. ELISA techniques are increasingly used for detecting parasite infection and the latest is for Chagas' disease. See also p. 194.

Immunoassay Kit Directory is a major publishing venture by Kluwer in, so far, three series. The first, called *Series A: Clinical Chemistry*, was published in four parts 1991 –. Each part is replaced every 12–18 months by a complete revision called a volume. Thus Volume 3, 1994–1995, will have five parts and cover more than 2000 kits. Each part has a review article of general interest. The contents are Peptide hormones (Part 1), Steroid and thyroid hormones (Part 2), Proteins and tumour markers (Part 3), Drugs, elcosanoids and second messengers (Part 4) and Equipment (Part 5). Part 3 includes complement, immunoglobulins, cytokines etc. *Series B: Infectious Diseases* consists of Volume 1 in three parts 1994–1995: Genitourinary infections (Part 1), Respiratory infections (Part 2) and Enteric, systemic and other infections (Part 3). *Series C: Immunology* 1995 will be in one or two parts. For each kit there details of the methods, format, sample type, reagents, controls, notes on interpretation (etc. set out in a uniform style for easy comparison. For each organism there is a short introduction. This publication should have wide appeal as new kits are constantly marketed together with changes and withdrawals. Further volumes are planned, including one for food and environmental subjects; eventually all the subjects may be brought together on a CD-ROM. There are perhaps about 12 000 kits on the market. The volumes are not cheap, but may be less than the cost of a single kit: comparison of the effectiveness of kits may be tested by quality control programmes (see p. 45 and 47).

Specialist topics

International Review of Immunology (quarterly) has each issue devoted to a single topic. *Immunodeficiency*, previously *Immunodeficiency Reviews*, is mainly reviews but now has some papers. Photocopying of these two journals and of *Vaccine* is not allowed, but VACCINE is now online.

Immunology has wide and important applications, for example in reproduction, neurology and development, for all of which there are specialist journals. Elsevier is a major publisher of immunology journals.

Clinical Immunology Newsletter (monthly) covers important topics such as monoclonals, cytokines, natural killer cells, suppressor cells and immunoassays.

International workshops and conferences for specialists play a very important role in immunology, and many are published and become reference works. *Leucocyte Typing Database* is a software package compiled by WR Gilks and sold by OUP. *Leucocyte Typing Database III* can be used to analyse the data from the Third International Workshop on Human Leucocyte Differentiation Antigens. Database IV, 1990, stores data from 1100 different monoclonal antibodies to 89 human leucocyte antigens.

Leucocyte Typing V by S.F. Schlossman (IRL, 1995) is the latest, with contributions from 500 laboratories. *HLA 1991. Proceedings of the Eleventh International Histocompatibility Workshop and Conference* (Oxford Science Publications) contains the results of the four-year international collaborative programme, with 537 sera and 32 000 panel cells from 180 ethnic groups.

The Leucocyte Antigen Facts Book, by A.N. Barclay *et al.* (Academic Press, 1993), very highly recommended by the staff of *Immunology Today*, includes the human leucocyte differentiation antigens from the fifth International Workshop, with the accession data for the various databases.

T-Cell Receptors, edited by J.I. Bell, M.J. Owen and E. Simpson (IRL, 1995) is a comprehensive review.

There are several databases and directories for hybridoma collections, e.g. through the MSDN and the catalogues of the ATCC (see p. 149). *Linscotts Directory of Immunological and Biological Reagents* lists hundreds of monoclonal antibodies and hybridomas against immunnoglobulins and their fragments and chains, cells and histocompatibility antigens, antigens such as complement, hormones etc., infectious agents, and hybridomas and myelomas as well as polyclonal antibodies against a wide range of antigens (see p. 44).

The Immunoclone Data Base (ICDB) is online as Immunoclones (IMMU) 1986– and is updated monthly with about 2000 additions. There is information on hybridomas, transfectomas, T-cell clones and their products and monoclonal antibodies. The descriptions give details of cell types, immunogen, immunocyte donors, immortal partners, reactivity, applications and patent or literature source. There are no data on polyclonal antibodies and kits. A very wide range of sources are searched, e.g. the online databases, patents and patent applications, collections from research laboratories, journal articles and catalogues. There is a printed monthly version, *Immunoclones*, which is also available on floppy disk.

Guidebook to Cytokines and their Receptors, edited by N.A. Nicola (OUP, 1994) is a necessary guide to this complicated subject. There is an appendix on chromosomal and genetic localization of genes for mouse and human cytokines and their receptors. *Advanced Immunology*, edited by D. Male, B. Champion, A. Cooke and M. Owen (2nd edn., Mosby, 1991) has an appendix on cytokines. *Cytokine Bulletin* is a free quarterly published R & D Systems (see also p. 164). The *FactsBook* series from Academic Press covers many aspects of immunology.

History of immunology

There has been a recent interest in the history of the subject with several symposia and books. The bicentenary of Jenner in 1996 and the centenary of Pasteur have been celebrated with meetings as far apart as London, Paris and Hong Kong: a major conference in December 1995 was published by Elsevier in 1996 in English, with a shorter version in French by Fayard: *Vaccinia, Vaccination, Vaccinology. Jenner, Pasteur and their successors*, edited by S. A. Plotkin and B. Fantini and *L'Aventure de la vaccination*, edited by A-M. Moulin. *A history of immunology* A.M. Silverstein, (AP 1989) is a readable history from 1720 to 1970. There is a short glossary, Nobel Prize highlights and a biographical dictionary. *Immunology 1930–1980. Essays on the History of Immunology* edited by P.M.H. Mazumdar (Toronto: Wall and Thompson 1989) is a collection of 21 essays from a symposium held in 1986. *Immunology: the making of a modern science* edited by R.B. Gallagher, J. Gilder, G.J.V. Nossal and G. Salvatore, AP 1995 traces developments since 1960 with 18 personal accounts by some of the chief scientists of this period. *Species and Specificity. An Interpretation of the History of Immunology* by P.M.H. Mazumdar (CUP 1995) is another interesting viewpoint of the central issue.

Postscript
Immunology Methods Manual, edited by I. Lefkovits, 4 vols (Academic Press 1996) has more than 2000 pages and is available as a CD-ROM with the usual facilities for bench use. *Immunoassy Automation*, edited by D. W. Chan comprises an original *Practical Guide* (AP 1992) and a supplement *An Updated Guide to Systems* (AP 1995).

Immunology Today had an Immune Receptor Supplement with four posters in 1996 and will publish a gene map of the human MHC in early 1997. Genzyme Diagnostics have a series of six coloured posters including one of cytokine receptor signalling issued in 1996. R & D Systems issued two posters on cytokines and one on CD antigens (up to date in November 1996), and a booklet *Fundamental Facts. Cell Adhesion Molecules*, edited by G.A. Wilson, 1996.

CHAPTER TWENTY-ONE

Parasitology

R. MULLER

In the last few years there has been an upsurge in interest in parasitology and studies on a nematode worm, *Caenorhabditis elegans*, albeit a free-living one, are at the heart of the ambitious international Human Genome Project. This upsurge has been stimulated by the new techniques of molecular biology, which are improving techniques of immunodiagnosis and are vastly increasing the feasibility of vaccines against parasitic diseases such as malaria and schistosomiasis; also by new and rigorous field studies which have demonstrated the clinical and public health importance of some common but often disregarded parasites of humans, e.g. the soil-transmitted nematodes and *Giardia*. Another fruitful field is that of the neurobiology of receptors in helminths. An exciting development is the campaign for the worldwide eradication of guinea worm disease, caused by *Dracunculus medinensis*, before the end of the decade. It has already been eliminated recently from Pakistan and almost from India, and there are intensive campaigns under way in West Africa. Progress is reported in the Guinea Worm Wrap-Up series produced by CDC (Centers for Disease Control), Atlanta.

Naturally, any increase in research work is reflected in the information produced; the state of knowledge in parasitology is moving very fast and frequent updating is necessary to keep abreast of current concepts.

Recent books and reviews

This list is by no means comprehensive, but is a personal recommendation of some of the recent books which illustrate various aspects of the subject.

Among general textbooks there are recent editions of well established textbooks such as *Parasitic Diseases*, edited by D.D. Despommier, R.W. Gwadz and P.J. Hotez, (3rd edn., Springer Verlag, 1994), notable for its outstanding lifecycle diagrams, the clearly written *Introduction to Animal Parasitology*, by J.D. Smyth, (3rd edn., CUP, 1994), the aptly named *Modern Parasitology*, edited by F.E.G. Cox (2nd edn., Blackwell, 1993) and the conventional but sound *Foundations of Parasitology*, by G.D. Schmidt and L.S. Roberts (Mosby, 1989). A recent and modern textbook is *A Functional Biology of Parasitism*, by G.W. Esch and J.C. Fernández (Chapman & Hall, 1993).

For medical parasitology there are the inexpensive and lavishly illustrated introductory text *Medical Parasitology*, by R. Muller and J.R. Baker (Lippincott, 1990), *Diagnostic Medical Parasitology*, by L.S. Garcia and D.A. Bruckner (Elsevier, 1988) for diagnosis, *New Strategies in Parasitology*, edited by P.W.J. McAdam, (Churchill Livingstone, 1989), *Parasitic Disease in Clinical Practice*, by G.C. Cook (Springer Verlag, 1990) and *Basic Clinical Parasitology* by F.A. Neva and H.W. Brown (Appleton and Lange, 1994) for clinical aspects of the subject, *Atlas of Human Parasitology*, (3rd edn., ASCP, 1990) and *Parasites in Human Tissues* (ASCP, 1995), both by L.R. Ash and T.C. Orihel and both beautifully illustrated, the really outstanding *Colour Atlas of Tropical Medicine and Parasitology*, by W. Peters and H.M. Gilles (3rd edn., Wolfe, 1989), *Atlas of Human Protozoa*, edited by E.G. Rondanelli and M. Scaglia (Masson, 1993), which is a more specialized atlas and the practical *Basic Laboratory Methods in Medical Parasitology*, produced by the World Health Organization (WHO, 1991). The relevant chapters in the magisterial and venerable *Mansons Tropical Diseases*, edited by G.C. Cook (20th edn., Saunders, 1996) are authoritative and up to date.

Veterinary Parasitology edited by G.M. Urquhart, J. Armour, J.L. Duncan, A.M. Dunn and F.W. Jennings (Churchill Livingstone, 1987) is still a standard textbook on this aspect of the subject and there is also a new sixth edition of another authoritative text *Georgis' Parasitology for Veterinarians* by D.D. Bowman (Saunders, 1995).

The best up to date treatises on pathology are the amply illustrated *Diagnostic Pathology of Parasitic Infections with Clinical Correlation*, by Y. Guttierez (Lea and Febiger, 1990), *Pathology of Infectious Diseases*, by F. von Lichtenberg (Raven Press, 1991) and *Atlas of Parasitic Pathology*, edited by K. Salfelder, T.R. Liscano and E. Saverleig (Kluwer, 1992), with numerous colour photographs.

There are plenty of books dealing with immunology and molecular biology, such as the comprehensive and authoritative *Immunology and Molecular Biology of Parasitic Infections*, edited by K.S. Warren (Blackwells, 1993), *The Biology of Parasitism: a Molecular and Immunological Approach*, edited by P.T. Englund and A. Sher (Alan R. Liss, 1988), *Genetics of Resistance to Bacterial and Parasitic Infection*, edited by D.M. Wakelin and J.M. Blackwell (Taylor and Francis, 1988), *Parasites: Immunity and Pathology*, edited by J.M. Behnke (Taylor and Francis, 1990), *Parasitic Nematodes – Antigens, Membranes and Genes*, edited by M.W. Kennedy (Taylor and Francis, 1991), *Modern Parasite Biology: Cellular, Immunological and Molecular Aspects*, edited by D.J. Wyler (Freeman, 1990), *Parasitic Infections and the Immune System*, edited by F. Kierszenbaum (AP, 1994), *Molecular Approaches to Parasitology*, by J.C. Boothroyd and R. Komoniecki (Wiley-Liss, 1995) and *Parasite Antigens, Parasite Genes. A Laboratory Manual for Molecular Parasitology*, by R.M. Maizels, M.L. Blaxter, B.D. Robertson and M.E. Selkirk (CUP, 1991). There is also a new textbook in German *Immunologische und Molekulare Parasitologie*, by M. Röllinghof and M. Rommel (Gustav Fischer, 1994). All these books provide accurate and up to date approaches to the subject.

On biochemistry there are a couple of good textbooks: *Biochemical Adaptations in Parasites*, by C. Bryant and C. Behm (Chapman & Hall, 1989) and the more specialized *Comparative Biochemistry of Parasitic Helminths*, edited by E.-M. Bennet, C. Behm and C. Bryant (Chapman & Hall, 1989).

Two still useful but not very recent books on chemotherapy are *Handbook of Drugs for Tropical Parasitic Infections*, by L.L. Gustafsson, B. Beerman and Y.A. Abdi (Alan Liss, 1987) and *Human Antiparasitic Drugs*, by D.M. James and H.M. Gilles (Wiley, 1985).

An outstanding and thought-provoking book on ecology and evolution is that by C. Combes entitled *Interactions Durables: Écologie et Évolution du Parasitisme* (Masson, 1995); unfortunately this book is not yet available in English. Other multi-author books in this field are *Parasite Communities: Patterns and Processes*, edited by G. Esch, A. Bush and J. Aho (Chapman & Hall, 1990), based on a symposium, and *Parasitic and Infectious Diseases: Epidemiology and Ecology*, edited by M.E. Scott and G. Smith (AP, 1994), which has some very stimulating chapters, particularly on mathematical modelling, as has *Ecology of Infectious Diseases in Natural Populations*, edited by B. Grenfell and A. Dobson (CUP, 1995).

More specialized recent and recommended books include: *Procedure Manual for the Diagnosis of Intestinal Parasites*, by D.L. Price (CRC, 1994), *The Physiology and Biochemistry of Cestodes*, by J.D. Smyth and D.P. McManus (CUP 1989), the comprehensive *Immunodiagnosis of Parasitic Diseases. I. Helminthic Diseases*, by K. Walls and P. Schantz (AP, 1986), *Parasitism and Host Behaviour*, edited by C.J. Barnard and J.M. Behnke (Taylor and Francis, 1990), *In vitro Methods for Parasitic Cultivation*, edited by A.E.R. Taylor and J.R.Baker (AP, 1987), another part of *Parasitic Protozoa*, edited by J.P.Kreier (vol 5, AP, 1993), *Nematode Parasites of Vertebrates: Their Development and Transmission*, by R.C. Anderson (CABI, 1992) and *Ecology of Marine Parasites*, by K. Rohde (CABI, 1993). Recent books on important human parasites include *Giardia: from Molecules to Disease*, edited by R.C.A. Thompson, J.A. Reynoldson and A.J. Lymberry (CABI, 1994), *The Impact of Helminth Infections on Human Nutrition*, by L.S. Stephenson (Taylor and Francis, 1990), *Echinococcus and Hydatid Disease*, edited by R.C.A. Thompson and A.J. Lymberry (CABI, 1995), *The Biology of Schistosomes: from Genes to Latrines*, edited by D. Rollinson and A.J.G. Simpson (AP, 1987), *Human Schistosomiasis*, edited by P. Jordan, G. Webbe and R.F. Sturrock (CABI, 1993), *Schistosomiasis: The St Lucia Project*, by P. Jordan (CUP, 1985), *Strongyloidiasis: a Major Roundworm Infection of Man*, edited by D.I. Grove (Taylor and Francis, 1989), *Hookworm Disease*, edited by G.A. Schad and K.S. Warren (Taylor and Francis, 1990) and *Ascariasis and its Prevention and Control*, edited by D.W.T. Crompton, M.C. Nesheim and Z.S. Pawlowski (Taylor and Francis, 1989), On the history of the subject there has been the indispensable and comprehensive *tour de force A History of Human Helminthology*, by D.G. Grove (CABI, 1990), the monumental *Catalogue of the Ross Archives*, by M.E. Gibson (LSHTM, 1983) and the *Ross Malaria Reprint* (The Keynes Press, 1988), and the computerized *Bibliography of Onchocerciasis*, by R. Muller and R.C.R. Horsburgh (CABI, 1987), which is published under keywords and authors and has a list of all the abstracts for each paper cited. Two personal records are the amusing and educative *Malaria Capers: more Tales of Parasites and People, Research and Reality*, by R.S. Desowitz (Norton, 1993) and *New Guinea Tapeworms and Jewish Grandmothers*, by R.S. Desowitz (Norton, 1981) and a book with a Canadian slant *Parasites, People and Progress: Historical Recollections*, by A.M. Fallis (Wall and Emerson, 1993).

The series *Advances in Parasitology* appears twice a year and includes five or six in-depth reviews on a wide range of current topics. *Parasitology Today* is bimonthly and has many shorter topical reviews, together with news, views, book reviews and lists of forthcoming conferences and courses.

The UNDP/World Bank/WHO Special Programme for Research and Training in Tropical Diseases (TDR) has been run by WHO in Geneva for over 20 years. Most of the diseases it has concentrated on are parasitic and include malaria, African and American trypanosomiasis, leishmaniasis, schistosomiasis, lymphatic filariasis (including dracunculiasis) and onchocerciasis. As well as funding research on these diseases, WHO produces a regular newsletter, *TDR News*, every four months and many practical publications. Recent ones include *Implementation of the Global Malaria Control Strategy* (1993), *Control of Chagas Disease* (1991), *Control of Schistosomiasis* (1993), *Lymphatic Filariasis: the Disease and its Control* (1992), *The Onchocerciasis Control Programme in West Africa* (E.M. Samba, 1994), *The Control of Onchocerciasis* (1995) and *Drugs for Parasitic Diseases* (in press). A list of electronic resources in tropical medicine is also available over the Internet (at ⟨walgate @who.ch⟩).

The most important human parasitic disease is malaria and there have been many recent advances in the treatment, diagnosis and control of this disease, [see *Malaria*, edited by A.J. Knell (OUP, 1991), *Malaria: Principles and Practice of Malariology*, by W.H. Wernsdorfer and I. McGregor (Churchill Livingstone, 1988) and *Essential Malariology*, by L.J. Bruce-Chwatt (2nd edn., Heinemann, 1985)]. There has been much recent research on malarial genetics and molecular biology and a vaccine is now being field tested (see C. Facer and M. Tanner, *Advances in Parasitology 39*, in press). New techniques of diagnosis include the use of acridine orange staining of malarial DNA (see T.E. Clendennen *et al.*, *Transactions of the Royal Society of Tropical Medicine and Hygiene* 1995; *89: 183*); a microtitre-plate hybridization (PCR) test with specific probes for the different species has been employed and there is a simple dipstick method based on monoclonal antibodies for the rapid diagnosis of *Plasmodium falciparum*, the most important species. However, for individual diagnosis the thick-stained blood film is still the routine choice. There has been an alarming spread of drug-resistant malaria in the last few years, and one of the new drugs synthesized to treat this condition is based on a traditional Chinese remedy [see *Chemotherapy and Drug Resistance in Malaria*, edited by W. Peters (2nd edn., AP, 1987)]. Concerning epidemiology and control of the disease one encouraging procedure is the use of impregnated bed nets [see P.L. Alonso *et al.*, *Lancet 1991:337; 1499* and *Mosquitoes, Malaria and Man: a History of the Hostilities since 1880,* by G. Harrison (John Murray, 1978)].

Primary journals

There are about 45 primarily parasitological journals in the world, 19 of which were listed in *Journal Citation Reports* in 1991. One recent trend has been an increase in publication in the English language, reflected in name changes for some long-established journals. Examples are *Parasitology Research*, formerly *Zeitschrift für Parasitenkunde, Parasite*, formerly *Journal de la Société Française de Parasitologie, Applied Parasitology*, formerly *Angewandte Parasitologie*, and *Tropical Medicine and Parasitology*, formerly *Tropenmedizin und Parasitologie.* In 1996 this last journal combined with three other journals to appear as *Tropical Medicine and International Health.*

Many parasitological journals have book reviews and some have short reviews outlining recent progress in various aspects of the subject.

Parasitology societies

Most countries or areas of the world have national or regional societies and in Europe, particularly following the demise of the Soviet Union, there have been a crop of new societies. In Armenia, the Baltic Countries, Czech Republic, Estonia, Georgia, Ireland, Russia, Slovak Republic and Ukraine, societies have recently affiliated with the European Federation of Parasitologists (Central Veterinary Institute, PO Box 65, 8200 AB Lelystad, The Netherlands). For details of parasitological societies worldwide reference can be made to the Secretary, World Federation of Parasitologists, PO Box 1, 3720 BA Bilthoven, The Netherlands. All societies have annual general meetings and scientific presentations, sometimes with published abstracts, and some have meetings on specialized topics which may be published.

Many societies have newsletters and bulletins: in the USA newsletters are also issued by regional societies. Most are available only to members, but some of the bigger societies are beginning to seek ISSN numbers and are thus becoming regular, accredited publications. The *Bulletin of the Royal Society of Tropical Medicine and Hygiene* and that of the American Society of Tropical Medicine and Hygiene, entitled *Tropical Medicine and Hygiene News,* are already in published form.

Electronic products

There are many projected sources of information for use on a microcomputer, but not many in production. These can be difficult to locate: for instance, the Liverpool School of Tropical Medicine/ WHO produce a very useful training manual on malaria microscopy on a floppy disk, but this is not advertised. There are also some CD-ROMs being produced, the most ambitious of which is the Wellcome Trust Tropical Diseases Videodisc Project. This will include CD-ROM, videodisc and video (this and other resources are produced by the Wellcome Centre for Medical Science, London NW1 2BE, UK). Nottingham University, UK, has produced a CD-ROM on biodiversity which has a parasitic component. Videos on parasitology have been produced in the UK by the Open University, the BBC and by pharmaceutical companies, e.g. Smith, Kline and Beecham, which can be contacted for details of hire.

Abstracting and online services and bulletin boards

Secondary series which provide abstracts on parasitology include the MEDLARS, BIOSIS, EMBASE and CABI databases (*Protozoological Abstracts, Helminthological Abstracts* and *Nematological Abstracts*), the *Microbiology Abstracts Section C* and the *Tropical Diseases Bulletin.* Online searches on MEDLARS have been greatly reduced in price and they have abolished user fees. For further details of these services see chapter 10. A comparison of secondary services in parasitology is given on page 55 and in Table 10.2, The first three databases are also available on CD-ROM, with frequent updates. A new CD-ROM has recently been produced by CABI on human health, CABhealthCD, including tropical medicine, and another is parasitology, CABparasiteCD.

ONCHO UPDATE and SCHISTO UPDATE were until recently published quarterly by the Edna McConnell Clark Foundation, 250 Park Avenue, New York NY 10177–0026, USA, and consist of abstracts of articles on onchocerciasis and schistosomiasis indexed in MEDLARS. They are available to interested workers on request. The former is being updated as a database on the Internet (at http://woodland.bio.ic.ac.uk/fgn/filgen1.html).

On the Internet there are many newsgroups, including those on bionet.parasitology (e.g. http://www.speakeasy.org/~dbrick/newspage/bionet.html and http://www.cs.ucl.ac.uk/newspage.bionet.html), bionet.immunology and bionet.protista.

Searching the World Wide Web (WWW) will disclose many user groups.

A complete bibliography of filariasis including onchoceriarsis is online at

http://helios.bto.ed.ac.uk/mbx/fgn/filgen1.htm.

A beautifully produced *Illustrated History of Tropical Diseases* is edited by F.E.G. Cox (Wellcome Trust, 1996) and includes many parasitic diseases of humans. *Malaria Research. An Audit of International Activity* by J. Anderson, M. MacLean and C. Davies (Wellcome Trust 1996) is an excellent review.

To join the discussion group on malaria organized by the Walter and Eliza Hall Institute in Australia you can e-mail listserv@wehi.edu.au with the message subscribe malaria and your e-mail number, name and address. Other groups available on malaria alone include: malaria page (cited in:

http://www.publiciastate.edu/entomology/resource list html);

malaria mouse strains

(http://www.wehi.edu.au/biology/malaria/who.html);

malaria antigen databases

(http:.ben.vub.ac.be/);

malaria databases

(http://golgi.harvard.edu/biopages/medicine.html;
http://www.primenet.com./~gwa/med.ed/specialties.html;
http://www.public.iastate.edu/~entomology/resourcelist.html,
and http://www.wehi.edu.au.biology.malaria.who.html);

malaria breakthrough

(http://www.sas.upenn.edu.african studies.articles gen.menu articles gen.html);

other malaria sites

(http://www.wehi.edu.au/biology/malaria/who/html).

There are similar groups for other parasites and special interest groups, and it would be pointless to give long lists of electronic addresses, particularly as they tend to change and are added to. One reference which could help to ascertain the current situation is the bionet parasitology index at the University of Oregon (http://darkwing.uoregon.edu/~joe/USENET/index.html). A directory of parasitologists worldwide with fax and e-mail addresses is being compiled by P.W. Pappas at the University of Ohio (e-mail pappas.3@osu.edu).

CHAPTER TWENTY-TWO

Zoology

A.W. HIBBERT AND N.J. ROBINSON

The diversity of animal species is almost matched by the amount of information available on them. Consequently, this chapter focuses on sources dealing with the core areas of zoology. The information compiled here is by no means totally comprehensive: the most useful sources are reviewed or references are given to other resources which can be used to access further information. After an introduction, the chapter is arranged by subject and then by animal group.

General sources

Zoological Record (ZR) is the oldest continuing and most comprehensive index to the world's zoological literature, covering around 70 000 references each year from over 6000 serials of over 100 countries. First produced in 1864 by zoologists associated with the Zoological Society of London (ZSL), it is now published jointly by BIOSIS and the ZSL. It is probably best known for its coverage of taxonomy and nomenclature, which is exhaustive, with each taxonomic change being separately recorded. However, the vast majority of entries deal with subjects outside these areas, such as ecology, behaviour, life cycle, methodology, anatomy etc. There are also references to conservation, meetings, conference proceedings, collections and museums.

Zoological Record is divided 27 separate books, one providing general references and 25 dealing with each major group, from protozoa to mammalia. Each book is arranged into author, subject, geographical, palaeontological and systematic indexes, so that references can be easily located by author, subject or animal name. Indexing is based on a hierarchical thesaurus, modified with free-text phrases which, in the absence of an abstract, convey the meaning of the item. The last book lists and references the new generic and subgeneric names included in the other sections of each volume.

Zoological Record is also available on CD-ROM and online (file 185 on the Dialog search system) from Volume 115, 1978–, holding over one million references in each format. These versions include a useful thesaurus of subject-controlled terms as well as taxonomic hierarchies. Support materials include the *Zoological Record Search Guide* and *Zoological Record Serial Sources*, the latter listing over 6000 serials from more than 100 countries covered in the *Zoological Record* database. This compares with only 100 listed under the heading of zoology in the *Journal Citation Reports* and the 500 or so core journals identified by using *Ulrich's International Periodicals Directory*.

General guides which may be useful as overall references include *A Guide to the Zoological Literature: the Animal Kingdom*, by G.H. Bell and D.B. Rhodes (Libraries Unlimited, 1994), *Using the Biological Literature. A Practical Guide*, by E.B. Davies and D. Schmidt, (Marcel Dekker, 1995), and *The History of Natural History: an Annotated Bibliography*, by G. Bridson (Garland, 1994), which gives an overview of all life science areas, including biographies and historical reviews.

General, well known life science bibliographic sources, such as *Aquatic Science and Fisheries Abstracts*, *Biological Abstracts*, *CAB International Abstracts*, *Science Citation Index*, *Current Contents* etc. are also vital sources for zoology, particularly in electronic form, where the larger databases are more easily searched and specific topics can be more easily isolated. Some providers produce subject specific subsets. (see Chapters 11 and 12).

There is an overlap between secondary sources but the degree of duplication differs between databases, and simply comparing journal coverage may not be enough to determine which database to use: different selection criteria are applied by different services, even for the same journal. Consequently, although two services may both index a given journal, only some of the items extracted may be duplicated (see Chapter 10 and Table 10.2). When selecting a service for bibliographic searches, the user needs to give equal consideration to the scope criteria employed and the level of literature coverage. It is unlikely that any single source will completely serve the information needs of the researcher. However, in practice the ultimate choice will probably depend just as much on availability as on content.

Information on the Internet falls into either primary or secondary information sources or links to other Internet sites. The latter are usually referenced, as they form data collections which are indexed and selectable on access. Individual uniform resource locators (URLs) have not been given as they change as systems are developed and enhanced. With the ever-increasing number of sites, the user is faced with the problem of locating the information. The Netscape browser is simple and adequate for most situations (available free via the Internet) and, used with the Lycos or Atla Vista search engine or the Yahoo subject index, it returns good results. It is essential to remember that data quality is not guaranteed, as it is possible for anyone with the equipment and knowledge to make data available. However, many reputable organizations have information on the Internet, and one can assume that their data has the same integrity as the normal services they provide.

The Internet also has useful general sites, including general guides: the *Biologists' Control Panel*, located at the Baylor College of Medicine, Texas, USA, and *Biologists' Guide to Internet Resources*, by Una Smith of Yale University. Publishers and book distributors like the Natural History Book Service and Specialist Scientific Books, UK, circulate their listings on WWW and orders may be made directly. There is a list of publishers' catalogues, organized geographically, and mounted by Northern Lights Internet Solutions Ltd of Saskatoon, Canada. Publishers of all disciplines are listed and the type of information appearing differs from catalogue to catalogue.

Many database providers, such as BIOSIS, also give information on their services and how to use them. The *Zoological Record* WWW and gopher site at BIOSIS UK, provides links to information sources, ranging from databases to jokes, pictures and sounds, arranged by animal group. Other links, arranged by subject, include natural history museums, zoos and publications (library catalogues, publishers' sites, journals etc.). This site also provides the names directory service for taxonomists (see below). The CAB International site gives information about its services and provides links to its Institute of Biological Control, Mycology Institute, Institute of Entomology and Institute of Parasitology.

Authored by Ken Boschert of Washington University, the *Electronic Zoo* lists Internet resources in zoology ranging from discussion lists and new groups to databases. *ZooNet* brings together all information on zoos throughout the world, with links to other animal-related pages. *NetVet* focuses on veterinary information and has many zoology links via its Animals section, including access to the *Electronic Zoo*. *BioNET-International* is a

Table 22.1. Source of published serials in Zoological Record 1994; 130 and 1996; 132 (from *Zoological Record Serial Sources* 5th edn., 1994 and 6th edn., 1996).

	1994	1996
Europe and Middle East	54.2	53.7%
North America	20.2	19.1%
Asia and Australasia	18.8	19.6%
Central and South America	3.1	4.1%
Africa	2.5	2.8%

general life science network accessed via the Internet. This is a global technical cooperation network dedicated to strengthening biosystematic capabilities and knowledge in developing countries. Dealing primarily with insects of agricultural significance, the network hopes to establish worldwide centres of biosystematic expertise to share information.

Zoological literature is not static: it is constantly changing as new publications emerge and existing ones fall into decline. Table 22.1 is a breakdown of the number of zoological serials by country of publication and illustrates the range of literature available. A similar representation would be true for non-serial publications.

General zoology texts usually adopt either a comparative form/function or a group/phylogenetic approach. Of the first group, *Zoology*, by R.L. Dorit, W.F. Walker and R.D. Barnes (Saunders, 1991) is typical. For the second, the two volumes of *Synopsis and Classification of Living Organisms*, edited by S.P. Parker (McGraw-Hill, 1982) is a classic text covering every animal group, many down to generic level. For a less weighty animal group approach, the *New Larousse Encyclopedia of Animal Life*, edited by M. Burton (revised edn., Hamlyn, 1989) provides broad coverage with excellent photographic illustrations. One noteworthy comprehensive reference work on zoology, published as a series of volumes on different animal groups, is the *Traité de Zoologie* series edited by P.P. Grasse, published irregularly and in random volume sequence by Masson from 1948 to the present day.

Although dictionaries are covered in Chapter 7, two examples are included here as they have been found to be particularly useful for zoology. *The Concise Oxford Dictionary of Zoology*, edited by M. Allaby (OUP, 1991) is a revision of the successful *Oxford Dictionary of Natural History*, with almost 6000 entries covering all animal taxa and additional subjects like evolution, behaviour, cytology, zoogeography, taxonomy and genetics. *The New Penguin Dictionary of Biology*, edited by M. Abercrombie *et al.* (8th edn.,

Penguin Books, 1990) is more general but still contains a wealth of zoological terminology.

The main journal publishers are Elsevier, Springer-Verlag, Academic Press, Nauka and Blackwell. For the majority of non-serial publications, including books, the main publishers are Academic Press, Springer-Verlag, Oxford University Press, CRC Press, Cambridge University Press, Nauka, Chapman & Hall, IUCN and Wiley. Interestingly, a large number of non-serial publications recently indexed in *Zoological Record* were published privately by the authors. Many publishers and booksellers produce subject-based lists of their stocks.

There are many libraries around the world which specialize in zoological literature. They can be located in various directories, e.g. *World Guide to Special Libraries*, edited by B. Bartz, R. Lochar and H. Opitz (3rd edn., 2 vols, Saur Verlag, 1990), which lists 161 libraries from 43 countries under zoology (see also Chapter 13). The *Bulletin Board for Libraries (BUBL)* site is hosted at the University of Bath, UK, and has an extensive subject tree with a zoology subdivision. There are links to other Internet sites, including directories, databases and much more, including a pilot project for a database of biologists.

Information on animal collections is available from various sources. The *International Zoo Yearbook* is published annually by the Zoological Society of London. Each issue covers a different type of zoo collection or zoo-related topic, e.g. the role of zoos in conservation. There are also special issues listing details of zoos throughout the world in a directory style. *International Index of Laboratory Animals*, edited by M.F.W. Festing (Laboratory Animals Ltd, 1992) has around 4000 entries of laboratory vertebrates, giving the names and location of stocks (see also p. 155). Large collections of type material are housed in many museums and universities around the world. Since 1978 over 800 items have been published on these collections in various books and journals (see p. 101). A most useful reference source for museums and universities is the book *World of Learning* (45th edn. Europa, 1995), which lists 26 000 museums, academic institutions and learned societies throughout the world, arranged by country. There is also a detailed index of institution names. Another useful directory is *Museums of the World*, by B. Bartz (5th edn., K.G. Saur, 1995), with information on location and collections of 24 000 museums worldwide. Museums are a useful source of information and many are now making their collection data available via Internet sites. The Smithsonian Institute Washington DC site focuses primarily on the museum and its publications, although there are some useful links to other organizations and discussion lists. Natural History Museum in London, UK, includes information on the museum but also gives access to its library catalogue. The *Peabody Museum* at Yale has a searchable archive of its collections in invertebrate zoology, invertebrate and vertebrate palaeontology, as well as herpetology, ichthyology, ornithology and mammalogy, with several hundred thousand items catalogued (see also p. 101).

Primary CD-ROM sources specific to zoology have emerged more recently, as technology has advanced and more exploitation has been made of multimedia. These discs usually include video and images as well as sound clips. The interfaces are not usually standardized and are of variable quality and ease of use. The *World Biodiversity Database CD-ROM Series* from the Expert Centre for Taxonomic Identification in The Netherlands is distributed by Springer-Verlag and UNESCO. So far, 11 discs have been produced: *Birds of Europe*, *Fishes of the North Eastern Atlantic and Mediterranean*, *Five Kingdoms: Life on Earth*, *Linnaeus Protist*, *Lobsters of the World*, *Marine Planarians of the World*, *North Australian Sea Cucumbers*, *Pelagic Molluscs of the World*, *Protoctist Glossary*, *Sea Mammals of the World* and *Turbellarian Families and Genera*. Each disc has full taxonomic descriptions and synonyms, hyperlinked text, distribution maps, photographs, video and sound clips.

Subject-based sources

Looking at the coverage of the zoological literature in *Zoological Record* since 1978, the consistent major subject is ecology (Fig. 22. 1) However, there has been a recent increase in systematics, particularly new taxa, which reflects the academic and public concern about biodiversity. Conservation and pollution are also attracting more attention, as green issues come to the fore.

Although animal behaviour is covered in Chapter 23, one excellent work is *Biology of Animal Behaviour*, by J.W. Grier and T. Burk (2nd edn., Mosby, 1992). Two good standard texts for physiology, with a functional approach, are the recent work *Animal Physiology*, by M. Stewart (Hodder & Stoughton and Open University 1991) and the classic text of the same title by K. Schmidt-Nielsen (4th edn., CUP, 1990). For a comprehensive guide to the physiology of all animal groups. *Comparative Animal Physiology* by P.C. Withers (Saunders, 1992) is recommended. R. McNeill Alexander is an authority on animal locomotion and therefore his general zoology textbook *Animals* (CUP, 1990) has a not surprisingly biomechanical perspective. Evolution is a subject

Fig. 22.1: Literature Breakdown by subject (from *Zoological Record on CD* 1978–).

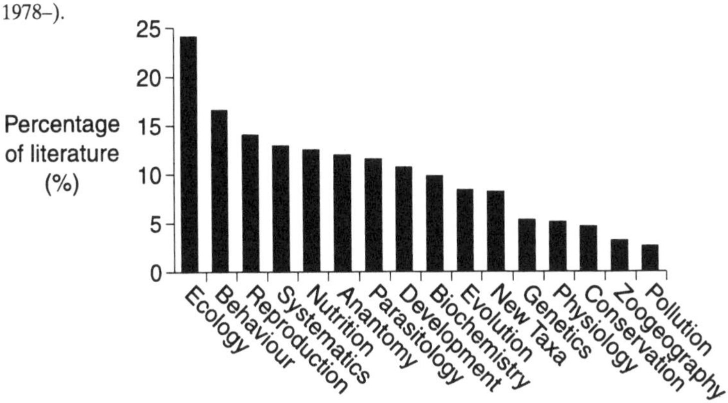

well served by the literature. Of the many, P. Skelton's textbook *Evolution. A Biological and Palaeontological Approach* (Open University, 1993) is a major reference source. This is a weighty tome by all standards, with over 1000 pages of information presented in a well organized and readable framework. A core textbook awaiting publication in the field of conservation is *Fundamentals of Conservation Biology*, by M.L. Hunter (Blackwell). For a more practical approach, D.L. Hawksworth's *Biodiversity Measurement and Estimation*, (Chapman & Hall and the Royal Society of London, 1995) is an informative reference.

Biodiversity

Much research has recently been put into biodiversity, which can be defined several ways. However, there is a resurgence in the area of taxonomy, with many organizations cataloguing the diversity of the world's fauna. CODATA and the International Union of Biological Sciences (IUBS) have cooperated to provide access to taxonomic data in the project *Species 2000*, which aims to enumerate all species of plants, animals and microbes as the baseline dataset for studies of global biodiversity by the year 2000. It is proposed that existing species databases and new systems will form a federation of datasets, with common access through media such as the Internet. CODATA and IUBS hope to catalyse the development of comprehensive databases by taxonomic experts in each subdiscipline of biology, and to assist biological database producers and users in gaining access to accurate taxonomic data. Further information about Species 2000 can be obtained from the Internet site. In another project, the White House Subcommittee on Biodiversity and Ecosystem Dynamics has also identified systematics as a research priority relating to ecosystem management and biodiversity conservation. Again, bodies such as the Interagency Taxonomic Information System (ITIS), a partnership of federal agencies, including the Smithsonian Institute and US Department of Agriculture, are working cooperatively with several similar international programmes to improve organization and access to standardized nomenclature. The goal is to create an easily accessible database with reliable information on species names and their classification, with the emphasis on North American taxa. For each scientific name the ITIS database will include the authority (author and date), taxonomic rank, associated synonyms and common names, and data quality indicator obtained by peer review. Further information and status reports on ITIS system development may be found at the ITIS Internet site.

Taxonomy

Principles of Systematic Zoology, by E. Mayr and P.D. Ashlock (2nd edn., McGraw-Hill, 1991) is the definitive textbook on zoological systematics. The 14 chapters cover theory, definition, classification, phenetics, cladistics, collections, taxonomic publication and interpretation of the rules of *The International Code of Zoological Nomenclature*. The latter aims 'to provide the maximum universality and continuity in scientific names of animals' and to give 'provisions and recommendations designed to enable zoologists to arrive at names for taxa that are correct under particular taxonomic circumstances.' It contains the valid rules of zoological nomenclature in an authoritative document, arbitrating on the validity of published taxonomic names. A draft of the fourth edition containing significant changes to the official recording of names is being circulated for comment and can be accessed via the *Zoological Record* WWW site. *Official Lists and indexes of Names and Works in Zoology*, edited by R.V. Melville and J.D.D. Smith (International Trust for Zoological Nomenclature, 1987) gives details of all the 9900 names and works on which the Commission has ruled since its inception in 1895. The latest update is the Second Supplement dealing with names added up to 1990.

For new taxa, Section 20 of the *Zoological Record* remains a primary source. It contains a list of newly described animal names

Table 22.2. Number of new generic and subgeneric names (from *Zoological Record* 1994; 130)

Animal group	Number of new names	Percentage of total new names
Smaller invertebrate groups	110	4.99
Protozoa	167	7.58
Coelenterata	41	1.86
Echinodermata	21	0.95
Vermes	128	5.81
Mollusca	167	7.58
Smaller arthropod groups (Trilobitomorpha, Arachnida, Arthropleurida, Merostomata, Myriapoda, Onychophora, Pentastomida, Pycnogonida (= Pantopoda), Tardigrada)	240	10.89
Crustacea	172	7.80
Insecta	970	44.01
Pisces	62	2.81
Amphibia	9	0.41
Reptilia	44	2.00
Aves	20	0.91
Mammalia	53	2.40
Total	2204	100.00

of subgenus level and above as part of its printed version (Table 22.2). In addition, the other sections dealing with individual animal groups provide details and bibliographic references to all taxonomic and nomenclature changes published in the 6000 serials and 1500 non-serial publications indexed.

The role of *Zoological Record* in taxonomy could be increased shortly if the discussion draft of the proposed fourth edition of the *International Code of Zoological Nomenclature*, due to take formal effect in January 1997, is adopted. The Code would then require that new names of species-group, genus-group and family-group taxa published after that date should be recorded in the *Zoological Record* within five years of their original publication, for them to be considered available. *Zoological Record* is investigating alternatives whereby taxonomists could check whether any new name published after 1996 has been recorded in the *Zoological Record* database. To obtain feedback from potential users a demonstration query facility, together with a mail form for comment, has been available since the summer of 1995.

Existing generic names can be found in *Nomenclator Zoologicus*, originally edited by S.A. Neave, which lists names published

Table 22.3. Breakdown of zoological literature by broad animal group (from *Zoological Record Serial Sources* 6th edn; 1996)

Animal group	Percentage of references
Vertebrates	41
Insects	23
Invertebrates (excluding arthropods)	20
Arthropods (excluding insects)	8
General zoology	8

in a given year or years with an indication of the broad classification and a reference to the description of the original type specimens. The series dates from 1939, with Volumes 1–5 covering 1758–1945. Following Neave's death Volumes 6–8 were edited by M.A. Edwards *et al.* and cover the period 1946–1977. Volume 9, also edited by Edwards, is in preparation and will cover taxa described in 1978–1994. All volumes are published by the Zoological Society of London. This is a valuable authority source for animal names until such time as a comprehensive electronic version is available.

Taxonomy also has a presence on the Internet. For three years starting in 1994, the European Science Foundation will fund a network for systematic biology in Europe. A moderated discussion list is based at the Natural History Museum in London to improve communication between European systematic biologists and their respective organizations. The list (SYSTEMATIC-BIOLOGY) distributes information on the activities and progress of the network, allows comment on the network's activities, and provides for exchange of information on systematic biology research across Europe.

Sources relating to animal groups

The main focus of this section is on taxonomic works and core monographs, rather than natural history accounts or popular works. Detailed coverage of parasitic organisms can be found in Chapter 21. By analysing the literature, it is possible to obtain figures for a breakdown of the references available for each major animal group (Tables 22.3, 22.4).

General invertebrate sources

Of all general invertebrate texts, perhaps the most widely cited classic is R.D. Barnes' *Invertebrate Zoology*, now in its sixth edition and coauthored by E.E. Ruppert (Saunders, 1994). Sadly, Barnes died before the book was published. A very useful anatomical reference work is *The Invertebrates: an Illustrated Glossary*, by M. Stachowitsch, (Wiley-Liss, 1992), with over 10 000 entries and descriptions of the major structural features of different invertebrate groups. Other general but comprehensive invertebrate tomes arranged by animal group are *Invertebrates*, by R.C. and G.J. Brusca (Sinauer Associates, 1990) *Invertebrate Zoology*, by P.A. Meglitsch and F.R. Schram (OUP, 1990) and *Invertebrate Zoology*, by J.A. Pechenik (2nd edn; W.C. Brown, 1991).

For fossil invertebrates the *Treatise on Invertebrate Paleontology* is a long-standing taxonomic authority. The volumes cover large and small taxonomic groups, for example Acritarcha, Conodonta, Mollusca and Trilobitomorpha. A useful general textbook is *Fossil Invertebrates*, edited by R.S. Boardman *et al.*, (Blackwell, 1987). It covers palaeontological topics such as the processes of fossilization, along with the biology of fossil groups and a chapter on trace fossils.

Specific animal groups

Protozoa – protozoans

There is no single comprehensive taxonomic authority for protozoans as they are such a diverse group. Some taxa appear in both zoological and botanical literature. The Society of Protozoologists Systematics Committee produced a stable and widely used classification scheme which was published by N.D. Levine *et al.* entitled 'A newly revised classification of the protozoa' in the *Journal of Protozoology* 1980: 27(1); 37–58. This provides a skeleton classification, with just a few representative genera included. Perhaps the most comprehensive textbook for all protozoan taxa is *Handbook of Protoctista*, edited by L. Margulis *et al.* (Jones & Bartlett, 1989). To complement this, Margulis *et al.* have recently published *Illustrated Glossary of the Protoctista* (Jones & Bartlett, 1993). Good general sources for abstracts are the *Cambridge Microbiology Abstracts C: Algology, Mycology and Protozoology* (see also p. 80).

The fossil groups seem to be more extensively documented, perhaps because of their geological associations with finite fuel

Table 22.4. Percentage of literature attributable to various animal groups (from references in *Zoological Record* 1994; 130)

Animal Group	Number of published references	Percentage of published references
General zoology	6207	8.33
Smaller invertebrate groups	628	0.84
Protozoa	4283	5.75
Coelenterata	725	0.97
Echinodermata	510	0.68
Vermes	4124	5.53
Mollusca	4033	5.41
Smaller arthropod groups (Trilobitomorpha, Arachnida, Arthropleurida, Merostomata, Myriapoda, Onychophora, Pentastomida, Pycnogonida (= Pantopoda), Tardigrada)	2673	3.59
Crustacea	3344	4.49
Insecta	17580	23.59
Pisces	8367	11.23
Amphibia	1409	1.89
Reptilia	2872	3.85
Aves	9497	12.74
Mammalia	8271	11.11
Total	74523	100.00

deposits. Good authority sources include H. Tappan's *The Paleobiology of Plant Protists (Freeman, 1990)*, 'Fossil dinoflagellates: index to genera and species' by J.K. Lentin and G.L. Williams in *A.A.S.P. Contributions Series* 1993: No. 28; 1–vii, 1–856 and A.R Loeblich and H. Tappan's two-volume *Foraminiferal Genera and their Classification* (Van Nostrand, 1988).

Of living taxa, a classic ciliate text is the comprehensive, but dated, work of J.O. Corliss, *The Ciliated Protozoa. Characterization, Classification and Guide to the Literature* (Pergamon, 1979).

Vermes – worms

B. Dawes' *The Trematoda* (C.U.P., 1968) despite its age, still remains a valuable and comprehensive taxonomic authority source for these platyhelminths. *Turbellaria of the World*, by L.R.G. Cannon (Queensland Museum, Brisbane, 1986) includes illustrated keys and diagnoses for taxa down to generic level and appears to be the only global authority for the Turbellaria.

Nematode taxonomic authorities include the four volumes of K.I. Skryabin's *Key to Parasitic Nematodes* (Amerind Publishing

Company and E.J. Brill, New Delhi, Leiden and New York, 1991–1992) and I. Andrassy's *Evolution as a Basis for Systematization of Nematodes* (Pitman, 1976). A useful, recently published, general textbook is *Nematodes. Structure, Development, Classification and Phylogeny*, by V.V. Malakhov (Smithsonian Institution Press, 1993). The *Caenorhabditis elegans Dataset*, a searchable bibliographic database including abstracts, at the University of Texas, can be accessed via the Internet. It has some advanced searching and linking features using the ACEDB search system. 'Nemertean genera and species of the world', by R. Gibson in *Journal of Natural History* 1995; 29(2): 27–562, is a unique annotated checklist which gives excellent cover. (See also p. 144)

Nomenclature Oligochaetologica. A Catalogue of Names. Descriptions and Type Specimens of the Oligochaeta and its later supplement, by J.W. Reynolds and D.G. Cook (University of New Brunswick, 1976 and 1981) provides an alphabetical list of all the genera and species of segmented worms up to 1981.

Mollusca – molluscs

There are numerous textbooks on molluscs, especially their shells. Of major taxonomic significance for the whole phylum is K.C. Vaught's *Classification of the Living Mollusca* (American Malacologists, Melborne, Florida, 1989). This work deals with all genus group names of living Mollusca, with subgenera, synonyms and author dates. There are also many catalogues which deal with a particular group or a particular geographical area, but only a few general sources are included here. The serial *Tryonia* consists of separate catalogues of molluscan species from different families or superfamilies. Many aspects of molluscan biology are covered by the 12 volumes entitled *The Mollusca*, edited by A. Wilbur and A. Willows (Academic Press 1983–1986). *Origin and Evolutionary Radiation of the Mollusca*, edited by J.D. Taylor (OUP, 1995) is a comprehensive textbook. A useful reference tool of over 1000 collectors and professionals is *The Directory of Conchologists/Malacologists*, edited by T. Rice, Of Sea and Shore Publications in 1993.

Smaller invertebrate groups, excluding arthropods

The simple sponges or Porifera can be particularly controversial in their taxonomy. The fossil species, for example the stromatoporoids, are the main problem. A comprehensive classification reference source for this group is C.W. Stearn's paper 'Classification of the Paleozoic stromatoporoids' in *Journal of Paleontology* 1980; **54**(5): 881–902. A recent taxonomic catalogue is 'Porifera' by J.N.A. Hooper and F. Wiedenmayeris in the series *Zoological Catalogue of Australia* 1994; **12**: i–xiii, 1–624, which covers sponge taxa, many with worldwide distribution, and is a well respected authority.

A recent general text for coelenterates is J.M. Shick's *A Functional Biology of Sea Anemones* (Routledge and Chapman & Hall, 1991). Similarly, P.J. Hayward *et al.* give the Bryozoa the same treatment in their book *Biology and Palaeobiology of Bryozoans* (Olsen & Olsen, Fredensborg, Denmark, 1994). A good taxonomic authority for entoprocts (sometimes considered to be bryozoans) is a German paper written by P. Emschermann in the serial *Marine Biology* 1972; **12**(3): 237–254.

For echinoderms, *A Functional Biology of Echinoderms*, by J.M. Lawrence (Chapman & Hall, 1987) is a useful general work, while 'Index of living and fossil echinoids' by P.M. Kier and M.H. Lawson in *Smithsonian Contributions to Paleobiology* 1978; No. 34: i–vi, 1–82, is a quality taxonomic reference.

Some of the minor enigmatic invertebrate groups are worth mentioning, as information can be difficult to find. A useful systematic overview of chaetognaths, or arrow worms, appears in *The Biology of Chaetognaths*, edited by Q. Bone *et al.* (OUP, 1991). The Sipuncula have a global catalogue down to species level written by J. Hylleberg and published in the *Phuket Marine Biological Center Research Bulletin* 1994; **58**: 1–88. The dated but unique taxonomic work of A.V. Ivanov, *Pogonophora* (Academic Press, 1963) lists most species and author dates.

Arthropoda – arthropods

Arthropod Name Index on CD ROM (ANI-CD) is a database of preferred taxonomic names of arthropods of agricultural, medical or veterinary importance from CAB International. It lists over 100 000 names, and functions include name checking, checklist production and viewing of taxonomic trees.

A good general text for crustaceans is F.R. Schram's *Crustacea* (OUP, 1986). It covers natural history, evolution and classification down to family level, including lists of families, for all major groups, with author dates and stratigraphical ranges appended. The only cautionary note is that the higher classification scheme used is peculiar to the author and frequently at variance with more traditional schemes. More specific taxonomic works include J.L. Barnard's two papers on gammaridean amphipods published in the journals *Bulletin of the United States National Museum* 1969; No. 271: i–vi, 1–535 and *Records of the Australian Museum Supplement* 1991; No. 13: 1–866. Also worthy of note for its vernacular name coverage is an article on decapods in the series on 'Common and scientific names of aquatic invertebrates from the United States and Canada', published in *American Fisheries Society Special Publication* 1989; **17**: i–vii, 1–77, by A.B. Williams *et al.*

For spiders, P.M. Brignoli's *A Catalogue of the Araneae Described Between 1940 and 1981* (Manchester UP, 1983) and N.I. Platnick's two updates, *Advances in Spider Taxonomy 1981–1987* (Manchester UP, 1989) and *Advances in Spider Taxonomy 1988–1991* (New York Entomological Society and American Museum of Natural History, 1993) are the major taxonomic authority sources. Mites are treated by S.W. Krantz in his systematic textbook *A Manual of Acarology* (2nd edn., Oregon State University Book Stores, 1978). R.L. Crawford's paper is also a good source for harvestman taxonomy in his catalogue from the journal *Burke Museum Contributions in Anthropology and Natural History* 1992; No. 8: 1–60. A relatively current general arachnid biology source book is *Spiders: an Illustrated Guide*, by R. Preston Mafham (Blandford Press, 1991).

For millipedes or diplopods, R.L. Hoffman's *Classification of the Diplopoda* (Geneva Museum d'Histoire Naturelle, 1979) gives global coverage from subclass to generic level with author dates throughout. *The biology of millipedes*, by S.P. Hopkin and H.J. Read (OUP, 1992) is a good general diplopod textbook.

Insecta – insects

The literature for the insect groups is overwhelming. This is not surprising, as the number of insect taxa far exceeds those of any other animal class. Consequently, P. Gilbert and C. Hamilton's *Entomology. A Guide to Information Sources* (2nd edn., Mansell, 1990) is an invaluable tool as it contains many of the classic serials and textbooks. There is also information on specimen collections, entomologists and organizations (see also p. 34).

Standard textbooks that have released later editions include the revised sixth edition of *An Introduction to the Study of Insects*, by D.J. Borror *et al.* (Saunders, 1989), the revised translation of B. Cymborowski's *Insect Endocrinology* (Elsevier, 1992) and *The Torre-Bueno Glossary of Entomology*, (New York Entomology Society and American Museum of Natural History, 1989), updated by S.A. Nicols and R.T. Schuh. M.T. Siva-Jothy's *Insect Biology: an Integrated Approach* (Blackwell, in press) includes an integrated approach to main subject disciplines such as evolution, physiology and behaviour. Other noteworthy textbooks published in the 1990s are *The Insects. An Outline of Entomology*, by P.J. Gullan and

P.S. Cranston (Chapman & Hall, 1994), *Aquatic Insects*, by D.D. Williams and B.W. Feltmate (CAB International, 1992) and *Insect Pathology*, by Y. Tanada and H.K. Kaya (Academic Press, 1993). ENTOMO-L is an e-mail notice board for communication between researchers, scientists and others in the field of entomology. The list is held at the University of Guelph, California, and has about 500 members. The Smithsonian Institute, Washington DC, also has a list of systematic entomologists at its Internet site (see also p. 101).

Coleoptera – beetles

R.A. Crowson's *The Biology of Coleoptera* (Academic Press, 1981) is still a good standard reference. In honour of Roy Crowson's 80th birthday a special selection of papers was published as a two-volume book in 1995. It is edited by J. Pakaluk and S.A. Slipinski and entitled *Biology, Phylogeny and Classification of Coleoptera* (Muzeum i Instytut Zoologii PAN, Poland). The serial *Coleopterist's Bulletin*, published four times a year, is a major source of new taxonomy, and the French book series *Les Coleoptères du Monde* is still producing major systematic revisions of beetle taxa. There are numerous taxonomic papers written on coleopterans but no overall definitive source, so a review of recent major revisions is given in Table 22.5.

Diptera – true flies

The order Diptera is well served by taxonomic catalogues. The 13 volumes of *Catalogue of Palaearctic Diptera* edited by A. Soos and L. Papp (Elsevier and Hungarian Natural History Museum, 1984–1993), give detailed family accounts, down to genus and below, with a final volume still to be published. R.W. Crosskey's catalogue in *British Museum (Natural History) Publication* 1980; No. 821: 1–1437, covers Afrotropical dipterans. Oriental taxa are catalogued in *A Catalog of the Diptera of the Oriental Region*, three volumes, by M.D. Delfinado and D.E. Hardy, published by the University Press of Hawaii 1973–1977. N.L. Evenhuis concentrates on the dipteran fauna of the Australasian and Oceanian regions in *Bishop Museum Special Publication* 1989; **86**: 1–1155. Finally, but by no means least, J.F. McAlpine's two-volume manual of Nearctic dipterans provides taxonomic coverage of this region in *Agriculture Canada Monograph* 1981; No. 27: i–vi, 1–674 and 1987; No. 28: i–vi, 675–1332. An interesting Internet resource is *FlyBase*, which deals with all molecular and genetic aspects of *Drosophila* (see p. 124, 144).

Table 22.5. Taxonomic revisions of coleopteran groups from journals, 1991 onwards

Taxon	Reference
Cerylonidae	Slipinski, S.A. *Annali del Museo Civico di Storia Naturale 'Giacomo Doria'* 1991; **88**: 1–273
Cetoniinae	Marais, E. and Holm, E. *Cimbebasia Memoir* 1992; No. 8: 1–125
Eucnemidae	Muona, J. *Entomologica Scandinavica Supplement* 1993; **44**: 1–133
Hydraenidae	Hansen, M. *Steenstrupia* 1991; **17**(1): 1–52
Hydrophiloidea	Hansen, M. *Kongelige Danske Videnskabernes Selskab Biologiske Skrifter* 1991; **40**: 1–367
Lucanidae	Maes, J.-M. *Revista Nicaraguense de Entomologia* 1992; Nos 22A and 22B: 1–121
Platydascillinae	Springer, C.A. and Goodrich, M.A. *Coleopterists Bulletin* 1994; **48**(1): 60–78
Platypodidae Scolytidae	Wood, S.L. and Bright, D.E. Jr *Great Basin Naturalist Memoirs* 1992; **13**: 1–1553

Lepidoptera – butterflies and moths

The classic taxonomic authority source for moths is the six volumes of 'The Generic Names of Moths of the World' I.W.B. Nye is editor of the five volumes by different authors, published in *British Museum (Natural History) Publication* 1975–1984; Nos. 770, 811, 812L, 848 and 880. The sixth, most recent, volume appears as a book, authored by Nye and Fletcher, published by the Natural History Museum, London, in 1991. All generic names are arranged alphabetically under the appropriate classification. For butterflies, F. Hemming's generic listing published in *Bulletin of the British Museum (Natural History) Entomology Supplement* No. 9: 1–509 is very thorough, although dated. There are many checklists and identification guides for this well photographed group. B. d'Abrera's *Butterflies of the World* series is full of colour plates and photographs. A good general and recent biology textbook is M.J. Scoble's *The Lepidoptera: Form, Function and Diversity*, (OUP, 1992).

Hymenoptera – ants, bees, wasps, sawflies etc.

G. Pagliano and P. Scaramozzino provide an overall taxonomic listing genera in 'Elenco dei generi di Hymenoptera del mondo' in *Memorie della Societa Entomologica Italiana* 1989; **68**: 1–210. For general textbooks, I. Gauld and B. Bolton's *The Hymenoptera* (British Museum (Natural History) and OUP, 1988) and B. Holldobler and E.O. Wilson's *The Ants* (Harvard UP, 1990) are standards.

B. Bolton is also the author of *A New General Catalogue of the Ants of the World* (Harvard UP, 1995) which provides more than 500 pages of detailed taxonomic information.)

Hemiptera – true bugs

A good standard text for this order is W.R. Dolling's *The Hemiptera* (OUP, 1991), which is well illustrated, with a glossary and section on information sources. Although the emphasis is on British species, the biology has a worldwide slant. Taxonomic catalogues include T. Henry and R.C. Froeschner's *Catalog of the Heteroptera, or True Bugs, of Canada and the Continental United States*, (E.J. Brill, Leiden, 1988) and three volumes of *Catalog of Hymenoptera in America North of Mexico* by K.V. Krombein *et al.* (Smithsonian Institution Press, 1979) *Leafhoppers (Cicadellidae): a Bibliography, Generic Check-list and Index to the World Literature 1956–1985*, by P.W. Oman *et al.* (CAB International, 1990) is also a useful work.

Smaller insect orders

Apart from the mass of literature on the largest five orders of insects, there is much written on the smaller, but not less significant, groups. A summary of taxonomic catalogues available for some of these taxa is given in Table 22.6.

General chordate/vertebrate sources

For fossil vertebrates the classic texts *Vertebrate Paleontology*, by A.S. Romer (3rd edn., University of Chicago Press, 1966) and *Vertebrate Paleontology and Evolution*, by R.L. Carroll (W.H. Freeman, 1988) are still pertinent. For living vertebrates there are several excellent alternatives to A.S. Romer's classic textbook *Vertebrate Body* (6th edn., Holt, R & W, 1986). The seventh edition of G.C. Kent's *Comparative Anatomy of the Vertebrates* (Mosby, 1992) is very comprehensive, as is R.G. Wolff's *Functional Chordate Anatomy* (D.C. Heath, 1991). New editions of other classics include the fourth editions of both *Vertebrates: Comparative: Anatomy, Function, Evolution*, by K.V. Kardong (W.C. Brown, 1995) and *Vertebrate Life*, by F.H. Plough *et al.* (4th edn., Prentice-Hall, 1996).

Table 22.6. Taxonomic catalogues for smaller insect groups

INSECT GROUP	CATALOGUE CONTENT	REFERENCE
Anoplura (sucking lice)	Worldwide species	L.A. Durden and G.G. Musser *Bulletin of the American Museum of Natural History* 1994; **218**: 1–90
Collembola (springtails)	Australian species	P. Greenslade in Houston and Greenslade *Zoological Catalogue of Australia* 1994; **22**: 1–184
Dermaptera (earwigs)	Worldwide genera	H. Steinmann *Series Entomologica (Dordrecht)* 1989; **43**: 1–934
Diplura	Australian species	W.W.K. Houston in Houston and Greenslade *Zoological Catalogue of Australia* 1994; **22**: 1–184
Ephemeroptera (mayflies) (see p. 28)	Worldwide genera	M.D. Hubbard *Flora & Fauna Handbook* 1990; **8**: i–viii, 1–119
Isoptera (termites)	Worldwide genera	M.J. Pearce and B.S. Waite *Sociobiology* 1994; **23**(3): 247–263
Mantodea (mantids)	Indian genera	T.K. Mukherjee, A.K. Hazra and A.K. Ghosh *Oriental Insects* 1995; **29**: 185–358
Megaloptera	Worldwide genera	J.D. Oswald and N.D. Penny *Occasional Papers California Academy of Sciences* 1991; **147**: 1–94
Neuroptera	Worldwide genera	J.D. Oswald and N.D. Penny *Occasional Papers California Academy of Sciences* 1991; **147**: 1–94
Odonata (dragonflies, damselflies)	Worldwide species	C.A. Bridges *Catalogue of family-group, genus-group and species-group names of the Odonata of the world*, Published by author, 502 W. Main Street, Urbana, Illinois, USA 61801, 1991
Orthoptera (grasshoppers, crickets)	Worldwide species	D. Otte *Orthoptera species file*, Orthopterists' Society and Academy of Natural Sciences of Philadelphia 1994–1995
Phasmida (stick insects, leaf insects)	Worldwide genera	P.E. Bragg *Phasmid Studies* 1992; **1**(2): 38–46 (see p. 29)
Protura	Australian species	W.W.K. Houston in Houston and Greenslade *Zoological Catalogue of Australia* 1994; **22**: 1–184
Psocoptera (booklice)	North American genera	E.L. Mockford *Flora & Fauna Handbook* 1993; **10**: i–xviii, 1–455
Raphidioptera	Worldwide genera	J.D. Oswald and N.D. Penny *Occasional Papers California Academy of Sciences* 1991; **147**: 1–94
Siphonaptera (fleas)	Valid worldwide genera	R.E. Lewis *Journal of Medical Entomology* 1993; **30**(1): 64–79

Specific chordate/vertebrate groups

Protochordata – Protochordates

This heading groups the Calcichordata, Hemichordata, Petalonamae, Urochordata and Cephalochordata, even though the first three taxa are often not included in the phylum Chordata. The amount of documentation on these groups seems to be inversely related to the degree of taxonomic controversy, i.e. the more controversial, the less documentation! Fossil taxa are better served than living taxa for taxonomic authorities in the guise of the *Treatise of Invertebrate Paleontology* series: inclusion in this series emphasizes the problematical nature of the classification of fossils from hemichordate groups such as Graptolithina, Enteropneusta and Pterobranchia. One of the few more comprehensive texts dealing with protochordate groups in any taxonomic detail is S.P. Parker's *Synopsis and Classification of Living Organisms* (McGraw-Hill, 1982). One other noteworthy source is a unique systematic review of all the species of the small phylum Phoronida by C.C. Emig in *Zeitschrift für Zoologische Systematik und Evolutionsforschung* 1974; 12(2): 128–151.

Urochordates or tunicates have more taxonomic documentation and are traditionally included in Chordata. The genera of Ascidiacea and the species of Appendiculata are catalogued by C. Monniot and F. Monniot in *Archives de Zoologie Expérimentale et Générale* 1972; **113**(3): 311–367, and R. Fenaux in *Mémoires de l'Institut Océanographique (Monaco)* 1993; No.17: I–VII, 1–123. Again, Parker's synopsis proves to be a useful taxonomic authority for cephalochordates.

Pisces – fish

Perhaps the most exciting book on fossil fishes to be published for a long time is the epic by J.A. Long entitled *The Rise of Fishes: 500 Million Years of Evolution* (Johns Hopkins UP, 1995). Not only are all the important genera from each group discussed, but hundreds of photographs and drawings are included as well. Its narrative covers the evolution of diversity derived from the fossil record of this large group. J.S. Nelson's *Fishes of the World* (3rd edn., Wiley, 1994) provides a good overall framework for classification for living and fossil taxa but only lists representative genera. The most comprehensive annotated generic list of living fish is W.N. Eschmeyer's *Catalog of Genera of Recent Fishes*, (Scientific Publications, California Academy of Sciences, 1990), arranged both alphabetically and hierarchically by classification. Of general works, P.B. Moyle and J.J. Cech's *Fishes: an Introduction to Ichthyology* (2nd edn., Prentice-Hall 1988) lives up to its name. Other recent broad textbooks are *Encyclopedia of fishes*, edited by J. Paxton and W.N. Eschmeyer (University of New South Wales Press, 1994), *Fish Ecology*, by R.J. Wootton (Blackie, 1992) and *The Physiology of Fishes*, edited by D.H. Evans (CRC Press, 1993).

FishBase: the fish CD, with key information on around 12 500 species of finfish and a user manual, is published by the International Centre for Living Aquatic Resource Management (see also p. 91).

Amphibia – amphibians

Two authoritative works for this relatively stable class are Frost's *Amphibian species of the world* (Allen Press and Association of Systematic Collections, Kansas, 1989) and W.E. Duellman and L. Treub's *Biology of Amphibians* (McGraw-Hill, 1986). The former is a world checklist of species with full type information. The latter is the classic textbook, which also includes the classification of Lissamphibia down to generic level, with author dates. Of new books advertised but not yet published, *A Natural History of Amphibians*, by R.C. Stebbins and N.W. Cohen (Princeton UP) and *Biology of Amphibians*, by B. Clarke (Chapman & Hall) appear to be good general sources. Amphibia and Reptilia are often treated together within the field of herpetology. As such, a publication which looks worthy of review is *A Complete Guide to Scientific and Common Names of Reptiles and Amphibians of the World*, by N. Frank and E. Ramus (in press). It will annotate over 10 000 species and their vernacular names.

Reptilia – reptiles

A primary taxonomic source for living reptiles is H.G. Dowling and W.E. Duellman's *Systematic Herpetology: a Synopsis of Families and Higher Categories* (HISS Publications, New York, 1978) which includes all the families and genera within a taxonomic hierarchy. Although very dated, A.S. Romer's textbook *The Osteology of the Reptiles*, (University of Chicago Press, 1956) is still a classic authority for the complete group, listing both fossil and extant genera (ignoring the obsolete classification scheme). A useful comprehensive biological overview is A. Bellairs and C. Cox's *Morphology and Biology of Reptiles* (Academic Press 1977). Of the mass of literature on separate reptile taxa a few salient works are cited here for major groups, with a global focus. *Australia's Reptiles on CD-ROM*, by S.K. Wilson and D.G. Knowles, is the compact disc version of the book *Australia's Reptiles*. Although restricted to Australian taxa, it contains over 600 colour photographs, distribution notes and descriptions. There are also some video clips, an entertaining trivia quiz and notes on the treatment of snakebite. Three well illustrated monographs on snakes are *Living Snakes of the World*, J.M. Mehrtens (Cassell, 1987). *The Atlas of Snakes of the World*, by J. Coborn (TFH Publications, 1991) and *Encyclopedia of Snakes*, by C. Mattison (Blandford Press, 1995). A good species account of alligators and crocodiles is S. Grenard's work *Handbook of Alligators and Crocodiles* (Krieger, Florida, 1991). Although based on data from the US Fish and Wildlife Service, it covers worldwide living species and subspecies, with a good bibliography. The biology of lizards is covered by the general textbook by C. Mattison, *Lizards of the World* (Blandford Press, 1989). Major chelonian works are *Turtles of the World*, by C.H. Ernst and R.W. Barbour (Smithsonian Institution Press, 1992) and P.C.H. Pritchard's *Encyclopedia of Turtles* (TFH Publications, New Jersey, 1979). The Internet *Sea Turtle Database* from the University of Florida contains a bibliography and is accessible by password via telnet.

The most popular of the fossil groups must be the dinosaurs, especially after the media spotlight from the film 'Jurassic Park'. Comprehensive textbooks *The Dinosauria*, edited by D.B. Weischampel *et al.* (University of California Press, 1990), *Dinosaurs: a Global View*, by S.J. Czerkas and S.A. Czerkas (Dragons World, London, 1990) and *Dinosaurs: the Textbook*, by S.G. Lucas (W.C. Brown, 1994) should satiate the curiosity of even the most avid fan.

Aves – birds

There are many detailed and reliable checklists and catalogues for this much studied class. Of them all, the second edition of R. Howard and A. Moore's *A Complete Checklist of the Birds of the World*, (Academic Press, 1991) is one of the most respected and more concise. The 16 volumes of J.L. Peters *et al.*, *Checklist of Birds of the World* (Museum of Comparative Zoology, Cambridge, Massachusetts, 1931–1987) is also a major reference. Other recently published checklists include the fourth edition of *Birds of the World: a Check list*, by J.F. Clements (Ibis Publishing, Vista, California, 1991).

Handbook of the Birds of Europe, the Middle East and North Africa, by S. Cramp *et al.* (OUP, 1977–1994 9 vols), and *Handbook of the Birds of the World*, by J. del Hoyo *et al.* (Lynx Edicions, Barcelona, 1992–) are major descriptive works.

A World Checklist of Birds, B.L. Monroe Jr and C.G. Sibley

(Yale UP, 1993) is drawn from the more detailed work, *Distribution and Taxonomy of the Birds of the World* (Yale UP, 1993). *World Checklist of Threatened Birds*, by J. Norton *et al.* (Nature Conservancy Council, England, 1990), has a conservation theme and is annotated with threat categories and other notes.

Bird identification is a very popular topic and there are very many field guides and faunal works on every region of the world in varying depths of detail. A recent and valuable reference tool to this literature is K. Adolfsson and S. Cherrug's *Bird Identification. A Reference Guide* (Sveriges Ornitologiska Forening, Stockholm, 1995). It provides over 10 000 references of birds identified in the western Palaearctic and Arabia.

A contemporary overview of avian biology is the second edition of F.B. Gill's textbook *Ornithology* (W.H. Freeman, 1994). *A Dictionary of Birds*, edited by B. Campbell and E. Lack (T & A.D. Poyser, Calton, England, 1985) is a weighty tome published for the British Ornithologists' Union.

A new CD-ROM is *Essential Ornithogical Abstracts*, which lists the contents and abstracts from six core ornithology journals from 1955 to the present. The journals covered are *Auk*, *Condor*, *Journal of Field Ornithology/Bird Banding*, *Ornithology Monographs* and *Wilson Bulletin*.

Mammalia – mammals

Mammal Species of the World. A Taxonomic and Geographic Reference, originally produced by J.H. Honacki, K.E. Kinman and J.W. Koeppl (Allen Press and the Association of Systematics Collections, 1982) has been revised and edited by D.E. Wilson and D.A. Reeder and reprinted by Smithsonian Institution Press, 1993. This key taxonomic authority is now updated with the 126 new species of mammals described since 1982. The other prime authority for mammals, *Walker's Mammals of the World*, edited by R.M. Nowak, (Johns Hopkins UP, 1991) is now in its fifth edition. These two volumes provide an illustrated catalogue of all the mammal species, with extensive coverage of vernacular names, population estimates and biology. A good standard biology textbook is T.A. Vaughan's *Mammalogy* (Saunders, 1986). Also recommended is D. Macdonald's *Encyclopedia of Mammals* (Hyman, 1989), which gives lots of biological details and is presented in a most readable manner, with photographs, sketches and maps to illuminate the text.

The *Multimedia Encyclopaedia of Mammalian Biology*, published by McGraw-Hill, is a searchable multimedia CD-ROM. Although the basic data are taken from the equivalent printed version, there are additional hyperlinks, sound and video clips. The disc also includes *Grzimek's Encyclopaedia of Mammals* (see p. 20).

Summary

It is impossible to do justice to the full range of literature. The emphasis of this chapter has therefore been general, and will hopefully provide a basis for further research. The aim is to guide, rather than to lecture. After all, literary criticism is, by nature, subjective.

CHAPTER TWENTY-THREE

Animal ecology and behaviour

T.D. WYATT

Introduction

Ecology is both a young subject and one of the oldest in biology. Although the roots of ecology can be traced in the writings of the ancient Greeks, for example Aristotle and Theophrastus, modern animal ecology had a slow start compared with plant ecology and it really grew after Charles Elton's classic 1927 book *Animal Ecology*, written in six months in his spare time. As Elton put it at the start of his book, 'ecology is a new name for a very old subject'.

Animal ecology and animal behaviour are natural partners and indeed, behavioural ecology, modern quantitative behaviour in an ecological setting, is established as a mainstream research area. Ecology and behaviour are developing rapidly as molecular techniques arrive and become better known both within and outside science. A behavioural ecologist, Nick Davies, tells of being caught up a tree watching (very) alternative mating strategies in dunnocks and being asked by a nun what he was doing. He explained, and to his surprise the nun replied that he should be using protein polymorphisms to determine who the parents really were (Davies, 1992). In the event, genetic fingerprinting provided even clearer answers.

This chapter will describe recent developments in animal ecology and behaviour and the way this has been reflected in the literature, and thus how to navigate the ecological web. As with other areas of biology, ecology is showing increasing specialization. One could imagine this to be perhaps partly driven by publishers looking for new journal titles, but it is also reflected in the growth of new scientific societies (which suggests a growth in specialization driven by scientists). As the subject grows new forums are needed.

The main journals

The pattern of journal types is much the same in ecology as in other subject disciplines. The journals with the most general and widest readership command the greatest kudos. So, for example, most ecologists crave an article in *Science* or *Nature*; relatively few of the papers in these journals are on ecology. Next in status are perhaps the more general ecological journals, such as *Ecology*, the *Journal of Animal Ecology* and *Oecologia*. More theoretical papers find a place in journals such as *American Naturalist*.

The trend for increasing specialization mentioned above has led to journals focusing on a particular taxon or habitat. For example, within animal ecology there is *Ecological Entomology* matched by *Mammal Review*, and of course long-running specialist bird journals such as *Auk*. Important animal species may have their journals, for example the domestic bee *Apis mellifera* has *Apidologie*. Divided by habitat, the journal *Freshwater Ecology* is matched by the *Journal of Arid Environments*. There is a similar trend, although less marked, within animal behaviour; for example, a recent new title is the *Journal of Insect Behavior*. How long before we have the *Journal of Freshwater Diptera Behaviour*?

New fields of ecology, such as molecular ecology and molecular evolution, also prompt new journals (see below). Here, as in behaviour, the techniques or research questions common to many projects tend to offer a unifying force to journals, and a wide range of species and habitats tends to be represented in each. In behavioural ecology, for example, it has also been trendy to work on both animals and plants, choosing the best species for the question or phenomenon under investigation. Equally, ecologists with molecular techniques can apply these to new taxa. One of the most useful current awareness journals is the monthly *Trends in Ecology and Evolution (TREE)* which features short refereed reviews on hot topics.

Reviews

These would be my first place to look when starting a new topic, but they tend to be densely written and do depend on a reasonable prior knowledge of the subject. The main reviews are those in the

Annual Reviews list (*of Entomology, of Systematics and Ecology*) (see Chapter 1). Other regular review series include *Advances in Ecological Research*, and in animal behaviour, *Perspectives in Ethology*. The symposia of the Zoological Society of the USA, published as grouped papers in the *American Zoologist*, form a useful series of reviews. Other major reviews can be found in the *Quarterly Review of Biology*, also a good source of book reviews of the main books recently published, sorted by subject, and *Biological Reviews*.

Reviews have the advantage that a subject and its key papers are put in context. The main drawback is that you are limited to the topics that have been chosen by others for review, and there may not be a recent review of the topic you wish to study. The second problem can be solved in part by using the *Science Citation Index (SCI)* to search for papers citing the older review. This may be your only solution if you are researching a topic not yet recognized in the literature, for example at the boundary of two disciplines.

Abstracts and databases

These are largely covered in Chapters 11 and 12. The division of ecology, in particular, and behaviour by habitat and taxon is a major handicap to finding literature, as papers about a topic could be published in a wide range of journals, from the specific to the general. Searching by keywords and forward searching from a key paper on the *SCI* are likely to be the best way to start. Most ecologists and animal behaviourists will find the main databases useful. However, given the limited journal cover of some of the database services and the relatively obscure journals in which nonetheless essential papers may be published, for example on the taxonomy or local distribution, it is worth searching more taxonomically based databases. These include *Ecological Abstracts, Zoological Record* (see Chapter 22), *Entomological Abstracts* (CABI) and others in that series. One must also not forget to search under plant headings in the case of herbivore–plant or habitat-based projects (see p. 84).

Animal ecology

Popular books on ecology and evolution have become best sellers in the 1990s. Although written for laypeople, many are soundly based and offer fluent, readable introductions to their topics. I include them here as they can be very useful even to experienced biologists. Notable among these books, and all available in paperback, are E.O. Wilson *The Diversity of Life* (Belknap Press, 1992), about the evolution of animals and plants in an ecological setting, M. Ridley *The Red Queen* (Penguin, 1993) on sexual selection and evolution, and J. Weiner's *The Beak of the Finch*, (Knopf, 1994), a masterly account of Peter and Rosemary Grant's remarkable study of Galapagos finches and the insights this gives on the process of evolution. I would also include R. Dawkins' books among these: *Selfish Gene* (OUP, 1989), *The Blind Watchmaker* (Penguin, 1991) and *Rivers of Eden* (Weidenfield & Nicolson, 1995). A very useful series of titles can be found in the *Scientific American Library* (Freeman), for example *Diversity and the Tropical Rain Forest* by J. Terborgh (1992). The classics include Charles Elton's 1927 *Animal Ecology* (Sidgwick & Jackson), which is still very readable and as a present day ecologist I have found many of his observations borne out by research since then. An account of the role of this influential scientist is given in P. Crowcroft *Elton's Ecologists: a History of the Bureau of Animal Population* (Chicago UP, 1991). Present-day textbooks are dominated by three North American authors: C.J. Krebs *Ecology: the Experimental Analysis of Distribution and Abundance* (Harper Collins, 1994), R.E. Ricklefs *Ecology* (Freeman, 1990) and *The Economy of Nature: a Textbook in Basic Ecology* (Freeman, 1993); and P. Colinvaux *Ecology 2* (Wiley, 1993). Another leading textbook is M. Begon, J.L. Harper and C.R. Townsend *Ecology: Individuals, Populations and Communities* (Blackwell, 1996).

Major publishers of more specialist ecology books, as well as textbooks, include the university presses of Cambridge, Oxford and Princeton, Blackwell, Academic Press, Sinauer and Chapman & Hall. Harvard and Chicago University Presses also publish significant numbers of ecology titles. Elsevier and Springer have large ecology lists. The reason for mentioning publishers rather than providing long lists of titles is that by the time you may be reading this, many more books will have been published. All of those mentioned have series with good reputations, although there are many good publishers not listed. Book reviews in *Nature, Science* and the *Quarterly Review of Biology*, as well as more specialist journals such as the *Journal of Animal Ecology, Journal of Biological Education* and *BioScience*, for example, may be the best way of proceeding.

The symposia of the British Ecological Society, published by Blackwell, are usually good starting points in the literature and not simply proceedings (see, for example, *Natural Enemies: the Population Biology of Predators, Parasites* edited by M.J. Crawley

(1992). It is worth mentioning again here that animal ecologists should be aware of the botanical literature even if they are not working specifically on animal–plant interactions (see Chapter 25).

Sampling, statistics and other techniques

A key part of ecology is the study of animal distribution and abundance. Many of the central techniques relate to this. *Ecological Methods* by T.R.E. Southwood (Chapman & Hall, 1978) remains a classic information source although it does focus on insect ecology. The best estimation of animal diversity has been a controversial topic. *Ecological Diversity and its Measurement* by A.E. Magurran (Croom Helm, 1988) provides a good introduction to practical approaches. *Ecological Experiments: Purpose, Design, and Execution* by G. Hairston (Cambridge UP, 1989) is one of a number of books on experimental design.

Two books can be strongly recommended for students carrying out projects: C.J. Barnard *et al. Asking Questions in Biology: Design, Analysis and Presentation in Practical Work* (Longman, 1993) and *The OU Project Guide*, by N. Chalmers and P. Parker (2nd edn., Field Studies Council, Preston Montford, UK, 1989). Both books are characterized by clear explanations and very practical advice. Barnard *et al.* is particularly strong on the design of hypotheses and the presentation of reports. *Animals of Sandy Shores*, by P.J. Hayward (Richmond Publishing, Slough, UK, 1994) is one of the latest of an excellent series of project-oriented books in a series of '*Naturalists' Handbooks*'. Although aimed at a primarily European readership, they provide a good introduction to each topic and, by highlighting the many things that are not known, encourage students to ask interesting questions.

Each researcher will have their own favourite statistics books. One of my own is R.R. Sokal and F.J. Rohlf, *Biometry* (3rd edn, Freeman, 1995). Given the large number of such books one approach may be to talk to a friendly statistician, although the best advice may be to find a book you find useful and have a number to hand, so that if you do not understand the theory in one you can try an alternative explanation which may be clearer. M.J. Crawley offers a practical approach to an upcoming technique in *An Introduction to Generalized Linear Models (GLIM) for Ecologists* (Blackwell, 1993).

Chemical and sensory ecology

For most animals the senses of taste and smell may be the most important, both for learning about the outside world and for communication within the species. The study of chemical ecology is a field which has started to blossom in the last 20 years. Much of the work is on pheromones, the chemicals animals use to communicate, but chemicals are also used for defence, by insects such as the bombardier beetle and the Monarch butterfly. A good overview of the subject is given by authors in *Chemical Ecology: the Chemistry of Biotic Interaction*; edited by T. Eisner and J. Meinwald (National Academy of Sciences, Washington DC, 1995). Chapman & Hall have perhaps the most extensive list of titles on pheromones and other aspects of chemical ecology; among these are *Insect Chemical Ecology: an Evolutionary Approach*, edited by B.D. Roitberg and M.B. Isman (1992) and *Chemical Ecology of Insects 2*, edited by R.T. Cardé and W.J. Bell (1995). The *Journal of Chemical Ecology* publishes a very eclectic mix of papers on topics ranging from hyena marking pheromones to defensive chemicals produced by marine organisms. Plants, which also use chemicals to suppress neighbours in interspecific interactions, are also well represented.

An excellent synthesis of the ways animals sense the world is given by *Sensory Ecology: how Organisms Acquire and Respond to Information*, by D.B. Dusenbery (Freeman, 1992). Two important journals are the *Journal of Experimental Biology* and *Comparative Biochemistry and Physiology A – Comparative Physiology.*

Aquatic ecology

Marine Biology: Function, Biodiversity, Ecology, by J.S. Levinton (Oxford UP, 1995) is likely to become a classic book on marine ecology, but *Marine Biology: an Ecological Approach*, by J.W. Nybakken (HarperCollins, 1993) is also a popular book now in its third edition. Both offer a synthesis across the subject. *Fundamentals of Aquatic Ecology*, edited by R.S.K Barnes and K.H. Mann (Blackwell, 1991) gives an overview of important topics, such as feeding in aquatic animals, looking for unifying principles across marine and freshwater ecology. Among textbooks in freshwater ecology M. Jeffries and D.H. Mills' *Freshwater Ecology: Principles and Applications* (Bellhaven, London, 1990) is a concise introduction. *Ecology of Freshwaters*, by B. Moss (2nd edn., Blackwell, 1988) is a more advanced text with an emphasis on man's impact on freshwater habitats.

Two important review serials are *Advances in Marine Biology* and *Oceanography and Marine Biology.*

Conservation ecology

Conservation biology is one of the fastest growing areas of ecology, and many new books have been produced in response. Among the leading textbooks are *Principles of Conservation Biology*, edited by G.K. Meffe and C.R. Carroll (Sinauer, 1994), *Essentials of Conservation Biology*, by R.B Primack (Sinauer, 1993) and *Biological Conservation*, by I.F. Spellerberg and S. Hardes (Cambridge UP, 1992). A useful series published by Wiley includes *Conservation in Progress*, edited by F.B. Goldsmith and A. Warren (1993). An enormous literature is developing about environmental impact assessment. One recent contribution is *Evaluation and Assessment for Conservation: Ecological Guidelines for Determining Priorities for Nature Conservation*, by J.F. Spellerberg (Chapman & Hall, 1992). See also p. 86.

Rehabilitation of habitats damaged by human activity is another active research area. One recent title is *Restoration Ecology: a Synthetic Approach to Ecological Research*, edited by W.R. Jordan, M.E. Gilpin and J.D. Aber (Cambridge UP, 1987). Two of the principal journals are *Biodiversity and Conservation* (1992–) and *Biological Conservation* (1968–). Debates about the ecological basis of conservation biology, for example about species–area relationships for nature reserves, appear in *TREE* from time to time. (see p. 5).

Molecular ecology

The application of molecular techniques has allowed new questions to be answered. In recent years the use of genetic fingerprinting to determine paternity (and thus mating patterns) in ecological studies has almost become routine, with examples in species as diverse as lions, wrens and pilot whales. Molecular techniques have also been used to study the dispersal of insects released for biological control. For molecular techniques in ecology there are a few books, for example *Molecular Genetic Ecology*, by A.R. Hoezel and G.A. Dover (IRL Press, 1991). More likely you will need to consult biochemists and other molecular biologists (see Chapter 18). Newsletters such as *Fingerprint News*, sadly now ended, offer an essential source of practical tips and wrinkles (see p. 28).

Behaviour

Animal behaviour is very popular subject with undergraduates and in response publishers have offered many titles. Among the best, with an evolutionary emphasis, are *Animal Behavior*, by J.R. Alcock (Sinuaer, 1993) and *Introduction to Behavioural Ecology*, by J.R. Krebs and N.B. Davies (Blackwell, 1993). *The Oxford*

Companion to Animal Behaviour, edited by D. McFarland (Oxford UP, 1981) remains a very useful overall introduction and reference guide. Animal behaviour does not have its own *Annual Reviews* series but many chapters in volumes such as those on *Entomology* and *Systematics and Ecology* may be relevant. One annual series worth checking is *Perspectives in Ethology*, edited by P. Knopfler and P.P.G. Bateson (Plenum Press) but the number of reviews in each volume is small. Volume 10 (1993) focused on behaviour and evolution. This is also the subject of a recent advanced text edited by P.J.B Slater and T.R. Halliday, *Behaviour and Evolution* (CUP, 1994).

In *Unravelling Animal Behaviour* (Longman, 1995), M.S. Dawkins gives a lucid introduction to the major topics perplexing researchers at present. She is particularly good at explaining the ambiguities behind such terms as 'communication'. An excellent introduction to studying behaviour in practice is given by *Measuring Behaviour: an Introductory Guide*, by P. Martin and P. Bateson (CUP, 1993). *The Study of Animal Behaviour*, by F. Huntingford (Chapman & Hall, 1984), discusses these issues from a more theoretical base. For guidance on the treatment of animals and the design of experiments to minimize the number of subjects, you could consult *Ethics in Research on Animal Behaviour*, edited by M.S. Dawkins and M. Gosling (Academic Press, 1992).

Companion animal behaviour

The scientific study of the behaviour of companion animals is relatively new, and it is ironic that we know relatively little about animals so close to our daily lives. The term 'companion animal' includes animals we would call pets, but only those which interact significantly with their owners (cats and dogs, principally, but not tropical fish). In some ways the impetus for studying companion animals came from the increasing difficulty of obtaining large grants to study, for example, lions in the Serengeti, so thoughts moved nearer home. At the same time the popularity of dogs and cats as pets in western countries has given rise to an enormous pet-supporting industry – from pet food to 'pet psychologists'. The popularity of pets makes them relatively easy to find and the business interest has provided some of the research funding. The subject is still establishing itself as an academic field. New academic societies, including the International Anthrozoology Society, which sponsors meetings and a newsletter, have recently started. There are an enormous number of popular books on the topic but few scientific ones. Books introducing the topic include J. Serpell's *In the Company of Animals* (Blackwell, 1986), which is particularly strong on pet keeping in non-western societies, among them South American native Americans, and *The Behaviour of the Domestic Cat*, by J.W.S. Bradshaw (CABI, 1992) and *The Domestic Dog: Its Evolution, Behaviour and Interactions with People*, edited by J. Serpell (Cambridge UP, 1995). Journals include *Anthrozoology.* Papers are also published in *Animal Behaviour* and other more general behaviour journals.

Animal welfare and the behaviour of domestic animals

Another important new area is the use of ethological techniques to provide more objective criteria for assessing animal welfare. The focus of research is on domestic animals such as battery hens, pigs and cows, but animals in zoos are also coming under study. Among the important topics are repetitive behaviour, such as pacing in zoo animals and 'mouthing' of cage bars by pregnant sows, and caging conditions in battery hens and other intensively farmed animals. As for companion animals, the literature is fairly young. Two recent titles are *Stress and Animal Welfare*, by D.M. Broom and K.G. Johnson, (Chapman & Hall, 1993) and *Farm Animal Behaviour and Welfare*, by A.F. Fraser and D.M. Broom (3rd edn., 1990 Baillière Tindall, London,). One of the main journals is the *Journal of Applied Ethology.*

Conclusions

Ecological and field behavioural research differs from much laboratory-based biology in some ways because much depends on local populations of animals at particular sites. Replication of studies is thus often difficult, and results are often published in local journals, which can make searching for them difficult. Nonetheless, most researchers are aiming for conclusions of wider relevance to ecology as a whole, and aim to publish in mainstream journals. Changes to allow the publication of large datasets of original data, associated with long-term studies in particular, may be one benefit of changes in electronic publishing and the use of the World Wide Web. Other trends in the subject are likely to mirror those in other fields outlined in this book.

Reference

Davies, N.B. (1992) *Dunnock Behaviour and Social Evolution.* Oxford University Press.

CHAPTER TWENTY-FOUR

Resources for plant sciences

H.V. WYATT

There has been a revival of interest in biological diversity as the disappearance of species throughout the world has highlighted the need for preservation. There are collections of many varieties, over 400 000 of wheat and 100 000 of maize, but there is considerable duplication – not necessarily a bad thing. The maizes in use are based on only 1% of the genus, with hundreds of hybrids derived from them. However, many of the hybrids are not unique for resistance to a particular disease or pest: a farmer using several different strains will be unaware that all may susceptible to a particular disease. The value of the study of systematics has been recognized by the formation of a UK Systematics Forum and a European Network for Systematic Biology to organize conferences, workshops and working groups. There are links with observers outside Europe: the groups are based on the Natural History Museum, London, UK.

Although botany is covered by *Biological Abstracts* and other general biological services, much information is also available in the literature of agriculture, forestry etc. The secondary serials and databases of these allied subjects should always be considered, especially as so much effort and money are spent on plants of economic importance (see p. 84).

The *Journal Citation Reports* lists about 120 journals under botany and about 100 under agriculture, perhaps the largest group of journals. Many of these are not mainstream and come from countries not in the usual North American and western European blocs. The methods of immunology, biochemistry and genetics are now common for all the sciences.

Guides

Guide to Information Sources in the Botanical Sciences by E.B. Davis (Libraries Unlimited, Littleton, Colorado, 1987) is a little old but still has excellent guides to floras, atlases, manuals, keys etc. and US and international learned societies. *Regnum Vegetabile* has occasional volumes which are valuable guides, e.g. volume 120 *Index Herbariorum. Part 1: the Herbaria of the World*; edited by P.K. Holmgren, N.H. Holmgren and L.C. Barnett (8th edn., 1990) published by the New York Botanic Gardens, includes an index to staff. Based on this volume is *Plant Specialists Index* by P.K. Holmgren and N.H. Holmgren (Koeltz Scientific Books, Konigstein, 1992) which has indexes to specialists in the systematics of plants and fungi throughout the world, with indexes under taxonomy, geography, subject, herbarium addresses, scientific names and specialists. Another guide in the *Regnum Vegetabile* series is *Taxonomic Literature: a Selective Guide to Botanical Publications and Collections with Dates, Commentaries and Types* by F.A. Stafleu and R.S. Cowen (2nd edn., 7 vols, Bohn, Scheltema and Holkema, 1976–1988). *Supplement 1* by F.A. Stafleu and E.A. Mennega (Koeltz Scientific Books, 1992) covers A-Ba. *The International Directory of Botanical Gardens* by C.A. Heywood, V.H. Heywood and P.W. Jackson (5th edn., Koeltz Scientific Books) lists 1400 gardens and arboreta with the taxa in cultivation, details of special collections, lists of seeds and seed banks, herbaria and publications. There is a geographical index.

A series of guides to the literature of the agricultural sciences is being published by Cornell University Press:

The Literature of Soil Science edited by P. McDonald (1994).
The Literature of Forestry and Agroforestry edited by P. McDonald and J. Lassoie (1995).
The Literature of Crop Science edited by W.C. Olsen (1994).
The Contemporary and Historical Literature of Food Science and Human Nutrition, edited by J. Brogdon and W.C. Olsen (1994).
Another guide is *Directory of Food and Nutrition Information*, by R.C. Frank and H.B. Irving (2nd edn., Oryx Press, Phoenix, Arizona, 1992).

Research centres and collections

There are many international groups and many databases and seed and germplasm collections. The International Board of Plant Genetic Resources, IBPGR, in Rome also controls the seed collection at the Vavilov Institute in St Petersburg, Russia, with 345 000 species. There are 13 International Agricultural Research Centres (IARC) with germplasm collections, research, libraries, publications etc. (see p. 104 for ICRISAT). The International Rice Research Institute (IRRI), Manila, for example, has more than 100 000 items on rice, of which 40% are publications from Japan. The Institute has special facilities for translation into English from Japanese, Chinese and Thai, and publishes *International Bibliography of Rice Research*, with supplements and a cumulative index 1950– and *International Bibliography on Cropping Systems*, with supplements 1973–. There are another ten international centres, including the International Center for Living Aquatic Resources, ICLARM, also in Manila (see also p. 91). There are more than 600 senior scientists from 50 countries working in these centres.

Collecting Plant Genetic Diversity: Technical Guidelines, edited by L. Guarino, V.R. Rao and R. Reid (CABI, 1995) is a massive compendium with a chapter on bibliographic databases for plant germplasm collectors. There is a Plant Genetics Resources Centre in Addis Ababa and a Centre for Genetic Resources in Holland. The European Union is involved in the systematic sequencing of new plant genes.

Kew Gardens

The premier centre in Britain is the Royal Botanic Gardens, Kew, known as Kew Gardens, which has a herbarium of six million specimens, a collection of 80 000 plant products, an arboretum in Sussex, an excellent library, a seed bank and many research activities. The *Kew Record of Taxonomic Literature* notes persons and botanical institutions with news items, biographies and obituaries, historical items with news of collections, new developments and descriptions of works. The library has over 200 000 illustrations, 500 portraits, 10 000 maps and 250 000 letters, as well as rare and valuable books. There is a list of booksellers, British and foreign, who handle specialized works in botany and horticulture. The library prepares extremely useful bibliographies. Kew has the world's largest collection of wild plants, nearly one in ten of all flowering species.

There is a Woody Plant catalogue and the Jodrell Laboratory identifies woods with their uses: probably the largest collection of wood specimens is held in the museum at Liverpool, a relic of the days when timber was imported through the docks. The Centre for Economic Botany in the Sir Joseph Banks Building at Kew has 75 000 items, and houses the Royal Pharmaceutical Society Herbarium and collection of crude drugs. The database has 40 000 items. There is an economic database for the 6000 species used for food, part of the Survey of Economic Plants for Arid and Semi-Arid Lands (SEPASAL) project, with a computer-based bibliography of economic botany. There are collections of economic botany, of which 70 000 items have been catalogued and photographed.

Book reviews and notices are published in the *Kew Bulletin*. *Herbarium News* has information about seed collections, publications, meetings etc. The Tropical Forest Research Group, based at Kew, has a worldwide databank of all research on tropical forests. *Kew Scientist* issued twice a year, gives details of meetings, publications, databases, e.g. Tree seed database, and research. The information is also on the Internet and the WWW.

Other British centres

The Botanic Gardens Conservation Secretariat, IUGN, is near Kew at The Green, TW9 3AA, UK. The World Conservation Monitoring Centre, University City, Cambridge, keeps a database of endangered species and habitats.

The Soil Survey and Land Research Centre, formerly the Soil Survey, at Rothamstead is now at Silsoe College, Milton Keynes, UK. Like many other institutes in Britain, it has passed from the public sector to an autonomous status: others have been closed, moved or amalgamated. There is continual change and it is difficult to keep pace with what is where. Many of the databases are associated with particular individuals and may move to new locations, with changes in their WWW address.

Various units of the UK Ministry of Agriculture, Fisheries and Food (MAFF) house collections: the Vegetable Gene Bank at Wellesbourne, the National Fruit Collection at Brogdale, the Pea Gene Bank at Norwich, the Commonwealth Potato Collection, the Barley Collection and the Soft Fruit Collection are all at the Scottish Research Institute at Invergowrie. There is a vegetable seed library and four seed sanctuaries at the Henry Doubleday Research Association, Ryton-on-Dunsmore, Coventry, Warwickshire, which supplies tree seeds and selects and evaluates potentially useful tree and shrub species for schemes abroad. The Cambridge University Botanic Garden publishes an annual Seed Exchange List and has a beans gene bank. A number of universities have gene banks of various species: fruit at Bristol, *Beta* and *Solanaceae* at Birmingham.

The database of the ecology of British vascular plants is at York, and, as the Ecological Flora of the British Isles, is online through BIDS as BIDS-ECOFLORA (see p. 69). The Checklist of the British

Flora is part of the BSBI database (see p. 250). A Central Taxonomic File is held at the Royal Horticultural Society, London. The Biological Records Centre at Monkswood has data on the distribution of both animals and plants. There is a central directory of all environmental data holdings for the UK at the National Environmental Research Council, NERC, at Swindon. A database of all UK-held datasets on global environmental change is kept by the UK Global Environmental Research Group at the Economic and Social Research Council, ESRC, also at Swindon. The Natural History Library of the BBC at Bristol has a database of all natural history films.

There are databanks of two national reference collections of garden plants at Leeds University. The Department of Agriculture, Northern Ireland, has flax and willow gene banks, and the NERC Unit of Comparative Plant Ecology has the pasture species gene bank. Units of the Biotechnology and Biological Sciences Research Council have gene banks of peas and cereals, forage grasses and legumes, vegetables including *Brassica*, *Allium* and *Daucus*, and fruit. The natural enemies of arthropod pests form the Natural Enemy Databank at the International Institute of Biological Control at Ascot. The *Viceae* gene bank, the Global Plant Species database and the Centre for Under Utilized Plants are also at the University of Southampton. For North American plants see p. 101.

For the UK there is a computer program, FLORA, for the accurate identification of 786 species of British wild plants, from Seale Hayne Faculty, Devon, TQ12 6NQ. There is a reference manual to illustrate botanical terms, but grasses, sedges, rushes and conifers are not included. In the UK many institutes publish through the government publishing agent HMSO, e.g. publications from the Institute of Terrestrial Ecology, which has seven research stations, and the Agricultural and Forestry institutes. These publications may be traced through SIGLE (p. 32) and the BLDSC publications (p. 33).

The International Organization for Plant Information, IOPI, at the Royal Botanic Gardens, Edinburgh, has a register of databases relating to plants and is compiling a checklist of the world's vascular species. At the same address are the European Taxonomic Floristic and Biosystematic Documentation System and the Flora Europaea database. *Guide to Standard Floras of the World*, by D.G. Frodin (CUP, 1984) is useful.

World centres

The world potato collection is at the International Potato Centre (CIP) at Lima, Peru, which issues the newsletter *CIP Circular* with gene maps, dendrograms and publications. There are other collections of wild species at Sturgeon Bay, Wisconsin, USA, and in Scotland (see above).

Crops of economic importance in developing countries are studied at IAR Centres (see above) and at many other institutes. The germplasm centre at Manaus, Brazil, holds 200 000 rubber genotypes and the Asian Vegetable Research Station in Taiwan holds soybean stocks. At ICRASAT there are germplasm gene banks of 10 000 pigeon peas, 20 000 of sorghum/millet and 10 000 groundnuts as well as minor millets. There is a *Pigeonpea Newsletter* of 50 pages a year with a circulation of 2500: grey literature and non-documented data are collected by staff, who travel widely. *The Pigeonpea*, edited by Y.L. Nen, S.D. Hall and V.K. Shula (CABI, 1990) was the third book in a series by the Consultative Group on International Agriculture Research. ICRSAT has an annotated bibliography and there is *CAB Prompts*, a current awareness bulletin. *Pulse Crops Newsletter* comes from the Indian Centre of Pulses Research, Kanpur, India. There is a *Chickpea Newsletter* with refereed and edited articles: 70% of the world chickpeas are grown in India. Details of some of the more important of these institutes are given in the annual *INASP Directory* (see p. 95).

There are a number of informal networks, including an Asian Grain Legumes Network. The International Legume Database and Information System, ILDIS, is online as *LegumeLine* from BIDS (p. 69). There are data on 18 000 species, subspecies and varieties, with distribution data down to state level. Taxonomic and botanical data are given, with references to published descriptions, illustrations and maps. The information is most complete for common species worldwide and all species from Africa, Europe and the Americas.

Peanut Research comes from Oklahoma State University, USA, and has a bibliography. *Bamboo Abstracts* comes from the Bamboo Information Centre, China. *Grass Literature*, a taxonomic newsletter from the Rijksherbarium, The Netherlands has a bibliography which is also available on disk. *Carnivorous Plant Newsletter*, despite the title, is a journal. The clonal germplasm repository system in the USA will have 12 facilities with the soybean and small grain collections at the NAL (see p. 100). There is a repository for fruit and nut plants at Corvallis, Oregon, and a lentil gene bank at the University of Saskatchewan, Canada. There are many newsletters.

The *Biotechnology in Agriculture* series 1990 – from CABI includes books on the potato, rice, barley etc. The latest is *Soybean: Genetics, Molecular Biology and Biotechnology*, edited by D.P.S. Verma and R.C. Shoemaker (1995).

Arabidopsis is an important plant for genetic study and there is a quarterly newsletter, an annual meeting and a stock centre for Europe at Nottingham University, UK, with mutants, DNA libraries etc. (see p. 00). *Methods in Arabidopsis Research*, edited by C. Konez, N-H. Chua and J. Schell (World Scientific, 1992) is a recent compendium of information. *Rhizobium* is another important organism used in genetic studies, with centres at USDA, Beltsville, USA, with 800 strains, and the MIRCEN centre in Brazil with 600 strains (see p. 137).

In Germany there are several centres with important worldwide collections and libraries, although some collections were lost in the war. The Botanischer Garten und Botanisches Museum Berlin-Dahlem and the Herbarium at Munich have two million specimens each; at Jena there are 2.5 million, with lesser collections at Halle, Gottingen and Hamburg.

In America the National Museum of Natural History, part of the Smithsonian Institution, Washington DC, has a herbarium of 4.5 million items, a fungus collection of 675 000 items and a collection of wood specimens (see also p. 101). The Missouri Botanic Gardens is almost as large. In Australia, the Royal Botanic Gardens of New South Wales and of Victoria have one million and two million specimens, respectively, and the Queensland Herbarium has half a million.

The ATCC has a catalogue of *Plant Viruses and Antisera*, with 800 items of viruses and viroids, molecularly cloned viruses and polyclonal antisera (see p. 150). ATCC also has a *Plant Tissue* list of its germplasm collection (see also under Guides, above).

Plant breeding

The traditional plant breeding techniques have been enriched by the newer techniques (see p. 41). *Molecular Breeding* 1995 – is a new journal about plants. Spr4inger has several new books, in what seems a major publishing drive. Two laboratory manuals are *Gene Transfer to Plants*, edited by I. Potrykus and G. Spangenberg (1995) and *Plant Cell, Tissue and Organ Culture – Fundamental Methods*, edited by O.L. Gamborg and G.C. Phillips (1995) (see also p. 130). *The Maize Handbook*, edited by M. Freeling and V.

Walbot eds (Springer, 1994) is very detailed, with protocols for molecular biology, genetics, cell culture, cell biology and development and morphology.

For sequenced plant genes there is the MENDEL database of the Commission for Plant Gene Nomenclature, CPGN, at Rutgers University, New Jersey, USA.

Secondary serials

Current Advances in Plant Science, culled from over 2000 journals, has a species index and gathers 26 000 references per annum. It includes symposia and reviews where these are given in a journal. There is a wide range of journals which are essentially experimental, e.g. in genetics, biochemistry and physiology. As with the other serials in the series, there are no subject indexes as such: the divisions point one to the subject areas and to 'other relevant titles' (see p. 77).

Biological Abstracts is used by many plant scientists (see p. 76) and there will useful information in many of the newer biotechnology secondary services where useful plants are concerned. *Current Contents/Agriculture, Biology and Environmental Sciences* is compiled from more than 950 journals (see p. 11).

The CABI database and the associated abstract journals are a major resource for the more agricultural side of the plant sciences (see p. 84 and 85). The specialist serials have their own useful features. *Seed Abstracts* has information about regulations. *Weed Abstracts* lists the common names, abbreviations and chemical names of herbicides and plant growth regulators (see also the Derwent databases p. 138).

Excerpta Botanica 1959 – Sectio A: Taxonomica et Chlorologia and Sectio B: Sociologica is a major abstracting serial for plant sciences. Unfortunately, in an era of cuts, this serial has suffered more than most. The BLSCD has not subscribed since 1987 when the large University of Leeds also ceased to buy it. Sectio A gives abstracts in several languages for articles on systematic botany, herbaria, gardens etc. Sectio B gives world cover of books and articles on plant geography and ecology. Volumes appear irregularly. Both sections are produced in conjunction with the International Association for Plant Taxonomy, Utrecht, Netherlands, which also publishes the *Regnum Vegetabile* series.

Among many publications from Kew are the *Index of Living Plant Collections in the British Isles* and the research newsletter *Botanic Gardens Micropropagation News 1991–*, with bibliographies, papers and research news.

Index Kewensis is the principal source for the names of Angiosperms. The Index gives the ranks and names together with the literature references. Looking through the Index and all the 19 Supplements is tiring. The CD-ROM contains the original 1893 edition plus the 19 supplements, and will be updated at intervals. A new Windows version, INDEX KEWENSIS 2.0, containing all updated data to the end of 1995 became available in January 1997. *The Kew Record of Taxonomic Literature Relating to Vascular Plants*, known simply as the *Kew Record*, indexes articles from 1300 journals, books and other publications with an annual author index and adds about 10 000 items a year. The main emphasis is on taxonomy, but there are also references to phytogeography, nomenclature, chromosome surveys, chemotaxonomy, floras and botanical institutions and other subjects, as well as bibliographies and biographies. The *Kew Record* is online and available from BIDS, with over 150 000 records, updated monthly (see p. 69). The equivalent for the *Bryophyta* is the *Index Muscorum* 5 vols 1959–1969 published as volumes of *Regnum Vegetabile*.

Abstracts in Mycology from BIOSIS has 20 000 abstracts per year whereas *Microbiology Abstracts Section C: Algology, Mycology and Protozoology* from the Life Sciences Collection, has only 10 000.

CHAPTER TWENTY-FIVE

Botany, plant ecology and the environment

H.J. PEAT

Botany and plant ecology

Interest in plants is ancient and references to them are found in many historic documents. Early interest was predominantly in their uses for food and medicine, but botany as a science emerged in the fourth century BC (Morton, 1981). Theophrastus (371–286 BC) published two works: *Historia Plantarum*, which covered the morphology and classification of plants, and *Causae Plantarum*, which dealt with plant physiology. Theophrastus not only included his own observations and experiments in these works, but also made use of existing scientific literature. The first journal for botanical science was the *Magazin für die Botanik* (1787–), but it was not until the 19th century that the subject really blossomed. The period saw advances in plant physiology, particularly in Germany, with the research and teaching of Julius Sachs, in plant geography as a result of increasing exploration of floras all round the globe, and in plant ecology. Plant ecology was not recognized as a distinct scientific subject until the end of the 19th century, with the publication of *Plantesamfund* (*Plant Ecology*), by E. Warming in 1895, although even the works of Theophrastus show evidence of ecological thinking as he recognized that species occur in particular habitats and that different regions tend to have distinct plant communities. The 20th century has seen a rapid increase in the scientific literature, not least in the botanical sciences. Much of the increase has been due to more specialized journals being published. In turn, this has led to more abstract and current awareness serials to aid the search for useful information. The latest important source of scientific information is the Internet and, in particular, the World Wide Web (see p. 17).

Primary sources

Botanical books and journals are published by a wide range of scientific and academic publishers, none of which particularly specialize in the plant sciences. Reviews of new books can be found in many major journals, such as *Annals of Botany*, *New Phytologist* and *Journal of Ecology*. For introductory texts the *Studies in Biology* series published by Edward Arnold and the *Outline Studies in Ecology* series published by Chapman & Hall include volumes on botanical topics. Recent texts on plant physiology are *Plant Physiology*, by F.B. Salisbury and C.W. Ross (4th edn., Wadsworth, 1992) and *Introduction to Plant Physiology*, by W.G. Hopkins (Wiley, 1995). Plant morphology is covered in *Plant Form: an Illustrated Guide to Flowering Plant Morphology*, by A.D. Bell (OUP, 1991), and aspects of plant ecology in *Plant Ecology*, edited by M.J. Crawley (Blackwell, 1986) and *Introduction to Plant Population Biology*, by J.W. Silvertown and J. Lovett Doust (Blackwell, 1993).

Books that either describe standard methods or critically review existing methods are extremely useful (see Chapter 8). Academic Press are producing a series, *Methods in Plant Biochemistry*, edited by P.M. Day and J.B. Harborne, with nine volumes published so far, each dealing with a different subject area. A wide range of ecological methods can be found in *Plant Physiological Ecology: Field Methods and Instrumentation*, edited by R.W. Pearcy *et al.* (Chapman & Hall, 1991) and *Methods in Plant Ecology*, edited by P.D. Moore and S.B. Chapman (2nd edn., Oxford, 1986) which reviews as well as describes relevant methods. Another useful text is *Methods in Comparative Plant Ecology: a Laboratory Manual*, edited by G.A.F. Hendry and J.P. Grime (Chapman & Hall, 1993) which is based on the methods used for the integrated screening programme of northern European plants by the Unit of Comparative Plant Ecology at the University of Sheffield, UK. One of the best ways to find out about a new subject area is to find a relevant review. A good review will discuss all avenues of research that have been made in the subject, and will include references to all important papers. The review journals *Botanical Review*, *Critical Reviews in Plant Sciences* and *Advances in Botanical Research* are devoted to plant research, but *Biological Reviews of the Cambridge Philosophical Society*, *Annual Review of Ecology and Systematics* and *Advances in Ecological Research* are also likely to contain articles of relevance to botanists and plant ecologists. In addition, there are also subject-specific review volumes, such as *Annual Review of Plant Physiology and Plant Molecular Biology* and review articles in some journals, e.g. the Tansley Review in *New Phytologist*, essay reviews in *Journal of Ecology*, and in the new monthly *Trends in Plant Science* 1996–. See also p. 70.

Journals

Almost 150 of the journals covered by *Current Contents* are entirely botanical in nature. These include journals which cover all areas, such as *Annals of Botany* 1887– and *New Phytologist* 1902– and an increasing number of more specialized journals such as the *Journal of Plant Growth Regulation* 1982–, *Molecular Plant – Microbe Interactions* 1988– and *Mycorrhiza* 1991–. Other journals focus on botany of interest to particular countries: America, The Netherlands, Bangladesh, Canada, Israel and Pakistan, for example, all have their own journals. Botanical information, however, is not restricted to these journals but may be found in many multidisciplinary journals, and any general journals may contain information of interest to botanists.

Few journals specialize in plant ecology, although *Journal of Ecology*, *Journal of Vegetation Science* and *Vegetatio* are notable exceptions. Much of research in plant ecology is published in ecological journals, such as *Ecology*, *Oecologia*, *Oikos*, *American Naturalist* and *Conservation Biology*, although general science and biological journals should not be forgotten. Most ecological journals can be found in *CC Agriculture, Biology and Environmental Sciences.*

Secondary sources

Secondary sources, therefore, are vital tools in seeking information that is of relevance to the individual botanist who does not have the time to scan the hundreds of journals which may, but usually will not, contain something of interest. *Current Advances in Plant Science* is ideal for keeping up to date with the recent botanical literature. This publication lists titles, bibliographic details and author addresses. The articles are ordered by subject under a number of main subject headings, with further subject divisions under each main heading. There are also author and subject indexes. An online version of this publication is available, but if

Table 25.1. A comparison of the contents and cover of *Ecological Abstracts* and *Ecology Abstracts*

	Ecological Abstracts	*Ecology Abstracts*
Publisher	Elsevier Science Ltd	Cambridge Scientific Abstracts
Frequency of issue	Monthly	Monthly
No. of abstracts 1994	c. 12 600	c. 12 100
No. of journals scanned	Over 600 journals, plus books, conference proceedings etc.	c. 180 core journals, plus over 6000 primary sources scanned
Information included	Bibliographic details, abstract (the author's where possible), translation of non-English language titles	Bibliographic details, abstracts, authors' addresses
Indexes	Subject, author, geographical, organism	Subject, author
Available media	Hardcopy, online on Geobase through several hosts, Silver Platter CD-ROM	Hardcopy, online through several hosts or using the Internet (see WWW page http://www.csa.com), Silver platter CD-ROM

it is being used mainly as a current awareness tool and relevant sections are scanned monthly then there are no major advantages in using an electronic rather than a printed version. Items of interest to the plant ecologist will also be found in the sister publication *Current Advances in Ecological and Environmental Sciences* (see p. 77).

There are no abstracting serials solely covering all areas of the plant sciences, so for a wide cover of the literature it is best to use *Biological Abstracts* (see p. 76). There are, however, more specialized abstracting serials covering particular areas. CABI publishes *Grassland and Forage Abstracts*, *Plant Breeding Abstracts*, *Review of Plant Pathology*, *Seed Abstracts* and *Weed Abstracts*. These are available online and on CD-ROM as part of CABI Abstracts (see p. 84) Of particular interest to British botanists is *BSBI Abstracts*, compiled by D.H. Kent and published annually by the Botanical Society of the British Isles, which includes bibliographic details and brief abstracts of articles and books on British vascular plants. The abstracts are arranged by subject, with sections for history, herbaria, biography and floras (ordered by geographical location) and also by taxonomy, making it very easy to search for the flora of a defined region or for a particular taxon. Its use is not restricted to British botanists as many of the species occur elsewhere, especially in northwestern Europe.

There are also abstracting serials for ecological information, although none specific to plants. Two useful examples are *Ecological Abstracts* and *Ecology Abstracts*. These are very similar, but *Ecological Abstracts* includes geographical and organism indexes, which are useful if the printed copy is used for a literature search (Table 25.1).

Electronic information sources

The Internet is becoming an important information resource in all kinds of areas, and especially in the scientific community, as it allows rapid, cheap communication between scientists and the chance to publish information speedily. Information on the Internet, however, must be looked at with even more critical eyes than printed information, as it is not subject to any kind of regulation or peer review. The sheer volume of material on the Internet makes searching for areas of interest a particular challenge. There are now numerous search engines available, such as the WWW Worm and Lycos, which allow keyword searching of World Wide Web documents.

A.R. Brach, of Missouri Botanical Garden and Harvard University Herbaria, has listed WWW sites of interest to botanists (Brach, 1995a): (*http://herb.biol.uregina.ca/liu/bio/botany.shtml*). Sites are listed alphabetically and a keyword search facility is included. He has also compiled a similar catalogue (Brach, 1995b) of sites of interest to ecologists: (*http://biomserv.univ – lyon1.fr/Ecology-WWW.html*), which is also ordered alphabetically with a keyword search facility.

For a catalogue organized by subject rather than organization, the one compiled by R. Lampinen at the Botanical Museum, Finnish Museum of Natural History, is useful: (*http//www.helsinki.fi/kmus/botmenu.html*). This covers categories such as arboreta and botanical gardens, botanical societies and organizations, conservation and threatened plants, databases, herbaria and university departments, and includes a brief description of many of the pages listed.

Another useful catalogue is the World Wide Web virtual library. This has pages for plant biology (*http://www.voknor.edu/cas/botany-micro/www-vl*) and biodiversity, ecology and the environment (*http://combio.bio.uci.edu/link/*), listing relevant WWW documents ordered alphabetically by organization and within each organization by subject.

There are an increasing number of databases holding botanical information as they are a convenient way to store large amounts of data, they can be kept up to date and the data can be easily searched and manipulated: ILDIS, International Legume Database and Information Service (Bisby, 1993), includes information on taxonomy, vegetation type, geographical distribution, conservation status and economic importance to man. Access is available through the Bath Information and Data Services (BIDS), details on *WWW* – *http://www.bids.ac.uk/ildis.html* (see pp. 69 and 242); Ecological Flora Database (Fitter and Peat, 1994) contains over 120 000 items of information on the ecology of 1777 British gymnosperms and angiosperms; 135 ecological characteristics are included covering habitat, distribution, morphology, physiology, life history and associated organisms. It is also available through BIDS (*http://www.bids.ac.uk/ecoflora.html*). PLANTS database of North American plants is produced at the National Plant Data Center. It includes information on taxonomy, geographical distribution, crop growth parameters, wetland indicator status and whether species are native or naturalized. It is intended to reflect the current state of knowledge from the botanical and agricultural communities and to help disseminate information to support conservation activities (details on WWW: *http://trident.ftc.nrcs.usda.gov:80/npdc/*).

Plant collections

Traditionally herbaria, botanical gardens and arboreta have been used for taxonomic and systematic work, but more recently their potential value for all kinds of research has been realized. Species in plant collections can easily be screened for compounds which may have pharmaceutical or agricultural uses, thereby avoiding the need for costly expeditions to find the species in their native environments. Historical herbarium specimens can be used for a variety of research; to look at changes in plant distributions (herbarium specimens may indicate that a species used to occur in localities where it is no longer found); or to measure the effect of man on the environment (earth adhering to herbarium specimens, for example, was the only source that could be found to calculate historic levels of heavy metals in order to assess the effect of testing nuclear devices in the Pacific (Pettitt, 1991)); or to investigate global change (such as comparing the stomatal density of living and herbarium leaves to look at the effects of rising levels of CO_2). Living plant collections such as botanical gardens and arboreta have an important role to play in plant conservation and the maintenance of genetic diversity (Heywood and Wyse Jackson, 1991). Finding which specimens are held where, however, is not always straightforward. Volumes in the *Regnum Vegetabile* series are useful: the location of the world's herbaria with a brief description of staff members, the collections and specific areas of research, can be found in Volume 120 (see p. 238) and the location of the collections of individual collectors can be found in *Index Herbariorum Part 2*, comprising Volumes 2, 9, 86, 93, 109, 118; information on botanical gardens can be found in Volume 95, *International Directory of Botanical Gardens*. It is less easy, however, to discover exactly which species from which localities individual herbaria hold, as very few have computerized records of their collections. Some of the world's major herbaria and botanical gardens are starting to computerize their collection records, particularly those of type specimens, and searchable catalogues are beginning to appear on the Internet. These may be found by using the botany catalogues described above, or by looking on the Biodiversity and Biological Collections Web Server produced at the University of Kansas Natural History Museum, which includes a botany section (*http://www.keil.ukans.edu/*). Once botanical collections likely to hold material of interest have been identified, the best way to find out exactly what is held is to contact the curator or to visit the collection (see also p. 43).

The environment

Awareness of the environment and man's effect on it has grown enormously over the last 20–30 years, and this has led to a huge growth in environmental research, literature and organizations. This ever-increasing wealth of information means that finding exactly the information you want is not always easy, therefore the aim of this section is to provide some starting points and some directions through the maze of environmental information.

A good place to look for basic environmental information, such as the meaning of a term, an explanation of an environmental issue or a piece of legislation, is in one of the many environmental encyclopaedias that have been compiled, such as *The Encyclopedia of the Environment*, edited by R.A. Eblen and W.R. Eblen (Houghton Mifflin, 1994). For a more in-depth discussion of environmental topics, such as acid rain, biodiversity, eutrophication or plant conservation, a useful publication is *Encyclopedia of Environmental Biology* edited by W.A. Nierenberg (3 vols, Academic Press, 1995), which covers over 150 topics and includes a brief bibliography for each. Another useful guide to environmental issues, especially for those concerned with North America, is *Environmental Studies: an Annotated Bibliography*, by D.M. Fortner (Metuchen, 1994). This is aimed at young people who want to discover more about environment issues and gives details and descriptions of readable books which do not require specialist knowledge to understand them.

Many of the academic, scientific and government publishers produce books and journals on environmental issues and concerns: Island Press, Washington DC, however, focuses exclusively on environmental issues and is a leading US publisher of such material. *The Island Press Bibliography of Environmental Literature*, compiled by J.A. Miller *et al.* (Island Press, 1993) is a useful bibliography, particularly of environmental literature in the 1980s. It is divided by subject, with extensive author, title and subject indexes, and a brief description of every item, including books and journal articles.

Publication of journals dedicated to the environment began in the late 1960s and new titles are still appearing, e.g. *Global Change Biology* 1995–. *CC Agriculture, Biology and Environmental Sciences* lists 150 journals in its environment/ecology section. Over a quarter of these have either environment or environmental in their title, and many others are concerned with environmental issues such as conservation, climate change, pollution and restoration.

There are many directories listing environmental information sources which cover primary and secondary literature sources, bibliographies, handbooks, databases, organizations and people. Table 25.2 gives details of the more up-to-date directories, with information on the types of sources included, the organization and the geographical coverage of each.

Secondary sources

Environment Abstracts, published by Congressional Information Service Inc., is available in printed form, on CD-ROM and through online hosts such as Enviroline. It aims to cover significant current research and analysis on the impact of mankind on the environment and on the effectiveness of remedial techniques and policies. There are 1000–1200 items abstracted monthly and over 800 journals systematically screened, which includes all major relevant English-language journals, including those published in third-world countries and a few foreign-language journals. Full bibliographic details are included, as well as author address, and the abstracts aim to summarize the objective, design/method of study, the groups affected, measurements made, the main results and the conclusions. The CD-ROM and online versions include over 240 000 citations and abstracts from *Environment Abstracts, Energy Information Abstracts* and *Acid Rain Abstracts*, dating from 1971 onwards.

Cambridge Scientific Abstracts produce monthly abstracting serials related to the environment, including *Agricultural and Environmental Biotechnology Abstracts* (3000 abstracts per year), *Aquatic Pollution and Environmental Quality* (7200 per year), *Ecology Abstracts* (12 100 per year), *EIS Digests of Environmental Impact Statements* (450 per year), *Pollution Abstracts* (11 000 per year) and *Toxicology Abstracts* (8800 per year). Together these form the Environmental Sciences database, containing over 700 000 records compiled from 7000 sources. This can be accessed using either online hosts or the Internet (see Table 25.1 and p. 78).

The environment on the Internet

There are a variety of catalogues on the World Wide Web dealing with the environment, of which the most extensive are listed below: WWW Virtual Library: Environment (http://ecosys.drdr.virginia.edu/Environment.html) is a catalogue of environmental sites listed alphabetically and by subject. The Virtual Environmental Library (http://envirolink.org/elib/) is another environmental catalogue produced by the EnviroLink Network as a clearinghouse of all environmental information available on the Internet. The catalogue is divided by subject, including areas such as environmental activism, government information and environmental issues, products and organizations. The environmental organization Web directory (http://www.webdirectory.com) began as an address database to keep environmentalists in touch and is now a large directory to sites in over 100 countries. It is divided by subject, and within each subject there are links to Web pages listed alphabetically by organization. There is also a list at the end of each subject category of the name and address of environmental organizations which do not have Web pages. Environmental acronyms (http://kaos.erin.gov.au/general/acronyms.html) is a searchable dictionary of acronyms produced by the Australian Environmental Resources Information Network, and although it has an Australian bias it includes acronyms of worldwide relevance. Eco-compass (http://www.islandpress.com/Eco-Compass/ECOCOMPHOME.HTM) is a monthly guide to environmental information on the Internet covering WWW, gopher and FTP sites, mailing lists and newsgroups. Environmental RouteNet (http://csa.com/routenet/access.html) is a subscription-based reference guide to environmental information on the WWW. In addition to pointers to relevant Web sites it includes access to 11 of Cambridge Scientific Abstracts' environment and pollution databases (see p. 78 and Table 25.1), an environmental news service (North American/international news) and newsletters from a variety of organizations (see also p. 27).

References

Bisby F.A. Species-diversity knowledge systems – the ILDIS prototype for legumes. *Annals of the New York Academy of Sciences* 1993; 70: 159–164.

Brach A.R. Botany and the Internet, useful and interesting World Wide Web servers – commentary. *Plant Science Bulletin of the Botanical Society of America* 1995a; 41: 7–8.

Brach A.R. Ecology and the World Wide Web – commentary. *Bulletin of the Ecological Society of America* 1995b; 76: 112–113.

Fitter A.H. and Peat H.J. The Ecological Flora Database. *Journal of Ecology* 1994; 82: 415–425.

Heywood V.H. and Wyse Jackson P.S. (eds) *Tropical Botanic Gardens: Their Role in Conservation and Development.* Academic Press, 1991.

Morton A.G. *History of Botanical Science.* Academic Press, 1981.

Pettitt, C. Putting 'bloody mice' to good use. *Museums Journal* 1991; 91: 25–28.

Table 25.2. Recently published directories and encyclopaedias on the environment

Source	Year published	Coverage	Organization of the information	Geographical region
Encyclopedia of Environmental Information Sources S. Balachandran ed, Gale Research	1993	Abstract & indexing services, bibliographies, directories, handbooks, online databases, periodicals, research centres	By subject	Worldwide
The Environment Encyclopedia and Directory, Europa	1994	Definition of c. 1000 terms, government and non-governmental organizations	Terms alphabetical, organizations by country	Worldwide
World Directory of Environmental Organizations by T.C. Trzyna and R. Childers 4th edn, Institute of Public Affairs, California	1992	Name and addresses of organizations	By country but also contains a list of key organizations under subject headings	Worldwide
European Environmental Information Sourcebook, compiled by O. Newman and A. Foster, Gale Research	1994	Organizations, books, directories, periodicals, research centres, databases	By subject	Most European countries
A directory of European Environmental Organizations by M. Deziron and L. Bailey 2nd edn, Blackwell	1993	Government and non-governmental organizations, giving the purposes, activities and publications for each one	Alphabetical	Europe
The Green Index – a Directory of Environmental Organizations in Britain and Ireland J.E. Milner, Cassell.	1994	Names and addresses of organizations	Alphabetical but with subject and keyword indexes	UK
ECO Directory of Environmental Databases in the United Kingdom M. Barlow, P. Fleming and J. Button eds, ECO Environment Education Trust, Bristol	1992	Description of environmental databases, including the type of information collected, information sources used, frequency of update, accessibility and contact details	Alphabetical by organization	UK
Directory of Environmental Consultants, 4th edn Environmental Data Services	1995	Lists companies dealing in environmental consultancy with information on the company, the areas of work covered and recent clients	Alphabetical	UK

Index